£20-00

Methane Production from Agricultural and Domestic Wastes

ENERGY FROM WASTES SERIES

Editor

ANDREW PORTEOUS

Reader in Engineering Mechanics
The Open University
Walton Hall, Milton Keynes, UK

Methane Production from Agricultural and Domestic Wastes

P. N. HOBSON, S. BOUSFIELD
and R. SUMMERS

Microbiology Department, Rowett Research Institute, Aberdeen, UK

APPLIED SCIENCE PUBLISHERS LTD
LONDON

APPLIED SCIENCE PUBLISHERS LTD
RIPPLE ROAD, BARKING, ESSEX, ENGLAND

British Library Cataloguing in Publication Data

Hobson, P N
    Methane production from agricultural and
    domestic wastes.—(Energy from wastes series).
    1. Methane   2. Sewage-sludge fuel
    3. Sewage sludge digestion
    4. Agricultural wastes—Recycling
    5. Gas-producers
    I. Title   II. Bousfield, S   III. Summers, R
    IV. Series
    665'.77   TP761.M4

    ISBN 0-85334-924-X

WITH 3 TABLES AND 25 ILLUSTRATIONS

© APPLIED SCIENCE PUBLISHERS LTD 1981

Printed in Great Britain by Galliard (Printers) Ltd, Great Yarmouth

Foreword

This volume in the Energy from Wastes Series covers the area of methane production from agricultural and domestic wastes. Principally this involves the conversion of excreta and other organic effluents to a valuable gaseous fuel plus, in many cases, a useful sludge for fertiliser or feedstuffs.

Dr Hobson and his colleagues have written a comprehensive text on the principles of microbiological processes and the biochemistry of anaerobic digestion, embracing the design of digesters with examples of current working installations. The potential for anaerobic digestion of wastes as diverse as sewage to fruit processing effluents is also reviewed.

This work should be of interest to all who have to manage organic waste treatment and disposal, as well as to a wider readership who wish to know more about methane production by anaerobic digestion.

ANDREW PORTEOUS

Preface

The production of methane, or more exactly, a flammable 'biogas' containing methane and carbon dioxide, by microbiological methods ('anaerobic digestion') is not new. The reactions have been in industrial use for over a hundred years, but only in sewage purification processes. In some times of national stress, such as war-time, the microbiological production of gas purely for fuel has been investigated, but with the resumption of plentiful supplies of fossil fuels the investigations have faded away. The idea that fossil fuels, particularly oil, were apparently able to supply man's needs made the search for other sources of energy not worthwhile, particularly as the other energy sources were seen to be more expensive than oil. But the growing awareness in the last few years that oil supplies will not last for ever, and that while many new sources of oil remain to be exploited the oil will be costly, has led to an increased interest in the production of energy from that inexhaustible source the sun, either directly or through the agency of the converters of sunlight into energy sources—the plants and animals of the earth.

These latter materials, 'biomass', can be converted into fuels by chemical and physical processes, but microbiological processes have the advantage of being possible on any scale and taking place at low temperatures and pressures. The microbiological process most near to widespread practical use is the production of biogas, and this book surveys the present state of the process.

Whereas a few years ago there were relatively few accounts of work on the subject, in the last few years the tempo of investigation has increased and many papers in journals and at symposia are devoted to the results of investigations. To quote them all would take a much bigger book than this. Some selection has been made, some work may have been overlooked, but the references given can lead the reader deeper into the subject. This is particularly so in the consideration of the microbiology of the process where only a brief review of the bacteria and their reactions is possible.

As a whole, the book tries to describe the basis of the biological and engineering problems involved in design and running of digesters and how laboratory and pilot-plant work is now being developed into full-scale plant for commercial production of biogas from many different kinds of organic matter.

The Aberdeen work on agricultural-waste digestion has been a collaboration between the Microbiology Department of the Rowett Research Institute and, initially, the Scottish Farm Buildings Investigation Unit and, latterly, the Engineering Division, of the North of Scotland College of Agriculture. The principal workers are named in the various papers quoted which describe the work, but Mr J. Clark took part in the initial experiments and Mr I. Auld and Mr D. Clouston have done the day-to-day work on the farm-scale digesters and kept records of their performance for many years.

P. N. HOBSON
S. BOUSFIELD
R. SUMMERS

Contents

CHAPTER 1

Introduction

As the scope of man's activities has increased, so has his need for energy. The simple needs of a primitive society, whether of long ago or today, can be met by the use of human or animal energy for mechanical tasks and by the use of wood or dried plant residues or animal excreta as a source of heat energy. And the needs of quite advanced civilisations can be also met with these simple energy sources: our own great cathedrals and castles, the cities of the ancient East and the South and Central Americas were built with human or animal energy alone. But life with these energy sources was slow; a cathedral or a pyramid took years to build and generations passed as some great cities took shape.

The invention and perfecting of the steam engine was the beginning of the increased pace of life in a large part of the earth and the beginning of the demand for large supplies of 'concentrated' energy. The engine could provide in a small space the energy of dozens of men or horses, but to do this it had in turn to be supplied with a source of this energy—something that could, by burning, supply large amounts of heat to generate steam. This source of energy was sometimes wood, but more generally coal, and so was developed a civilisation based on coal as a primary energy source. It was also found that coal could be turned into another source of energy, coal gas, and the era of gas lights and gas engines began. The steam engine and later, the steam turbine, could also generate the newly discovered electricity in large amounts and electricity superseded gas as a lighting agent, and it could also be used to provide mechanical energy through electric motors.

However, although steam engines and turbines could be made very powerful, their use was limited by the large amount of the primary energy source that they required, and this limitation was particularly marked in the supply of energy for movement of people and goods. A ship could carry hundreds of tons of coal and undertake journeys of thousands of miles and days or weeks in extent. A land vehicle, such as a tractor or lorry, was limited to a ton or so and could go only a few miles without refuelling. Even a railway locomotive could go for only a short time without a refuel, although it could carry some tons of coal. And although some steam-powered models were made, a full-size aircraft driven by steam was an impossibility, largely because of the weight of fuel needed.

1

Then came oil for lighting and the invention of the internal combustion engine. Because the burning of the fuel was the actual driving force of the engine, this did away with the cumbersome boiler of the steam engine, and the fuel was lighter and easier to handle than coal. Thus lighter and faster land vehicles were made and powered aircraft were possible. Based on oil have developed the transport systems of the present day, and oil has replaced coal as an energy source in many static systems, such as the electricity-generating station.

However, resources of oil and natural gas are finite and although there are coal reserves to last for hundreds of years, supplies of these fossil fuels will become scarcer. These fuels came originally from plant and animal life and while the processes that made them ceased millions of years ago, plants and animals still exist and continually renew themselves. Although the processes that led to coal and oil formation are still going on to some extent as life forms die and decay, the processes cover such a time span that the reserves of coal and oil are virtually non-renewable; indeed, the geological conditions which led to the formation of vast coal and oil fields may never occur again. Even peat, which is a form of decayed vegetation probably formed over a much shorter period than the coal–oil measures is non-renewable.

Vegetation and animals were the source of the fossil fuels, but the only way in which they can now form energy sources is when used immediately, without the aeons of decay and transformation in the earth which led to the fossil fuels. Because vegetation and animal life renews itself month by month and year by year it would seem to form an inexhaustible supply of energy. But it has drawbacks. Burning is the only way of obtaining energy directly from vegetation, but only in limited circumstances will it burn, and as was said before it is bulky and unsuited to most modern energy requirements. Only after some form of conversion process will freshly dead vegetation and animals give fuels which can be used for modern energy requirements, and forecasts of shortages of fossil fuels have revitalised research into ways in which such conversions can be carried out.

While calculations of the total biomass available on earth show that there are vast amounts and that these could in theory yield a large proportion, or even all, of the energy requirements of man, in practice there are many reasons why this full potential can never be exploited. This book deals with one method of converting renewable biomass to a 'concentrated' and easily utilisable source of energy, a gas, but it does not suggest that this could be a major source of world energy, the fossil fuels and atomic power will have to be that in the near future. The authors see the process and the gas as an additive to present energy sources and a substitute for these in certain circumstances. No attempt is made to calculate the total masses of feedstock available or the gas that could be generated if all this were used. At the present stage of development such figures are purely theoretical and largely meaningless. Developmental work needs to be done with full-scale plants to solve engineering problems, but the economics will take care of themselves. As energy increases in cost and scarcity, what were once processes that were not economically viable when set against the cost of plentiful and cheap oil become economically worthwhile.

CHAPTER 2

Methods of Production of Fuels from Biomass

'Biomass' can be described as organic material, organic in the sense of being, or being obtained from, living organisms; be these macro-vegetation and macro-organisms, or micro-organisms. The biomass contains carbohydrates, proteins, lipids, nucleic acids, non-protein nitrogenous compounds, and salts, but to obtain useful energy-containing compounds these have to be converted into a limited number of compounds of carbon, hydrogen and oxygen—carbon itself, carbon monoxide, hydrogen, methane or higher gaseous or liquid paraffins and olefins and alcohols. The nitrogen and sulphur in proteins and other compounds are, so far as energy production is concerned, only a source of impurities in the fuels which can eventually lead to problems of pollution or corrosion in the use of the fuels.

There are two methods of producing fuels from biomass; one involves the use of physical or chemical processes at high temperature and/or pressure, the other involves the use of micro-organisms at low temperature and atmospheric pressure. Although in some cases a chemical process is a preliminary to a microbiological one, the chemical processes involved do not call for, at the most, temperatures and pressure beyond those of wet steam.

There are two other differences between the physical and biological processes in that whilst dry material can be used for physical processes, and is essential in some cases, the biological reactions must always be carried out on material that is at least damp or more generally very wet or in solution or suspension in water. However, although a wet feedstock carries penalties in the form of a large proportion of inert material, most biomass is wet; even apparently solid vegetable and animal structures are some 80 % water. So the fact that microbiological processes need a wet feedstock makes them more generally applicable than the physical processes which need a dry feedstock.

PHYSICAL AND PHYSICO-CHEMICAL METHODS

The direct way of producing energy from biomass is by burning it in the presence of excess air and so oxidising the carbon- and hydrogen-containing compounds to carbon dioxide and water. Whilst this is quite possible with some dry materials such as woods, straws and other crop residues, and materials such as paper, made from vegetation, in general the substances are not good fuels in that their density is low and their calorific value is low compared with their volume. Such materials are used, though, particularly on sites where they are a waste from some manufacturing process and their availability and low cost outweigh their disadvantages as fuels. An example is bagasse, the sugar-cane residue, which is used as fuel in factories producing sugar from the cane.

The indirect way of producing energy from biomass is to convert it into a fuel of high calorific value which can be more readily transported or otherwise distributed. This means producing a liquid or a gaseous fuel. Methods of doing this by physical or physico-chemical processes involve either pyrolysis or hydrogenation. These methods and combustion processes will be described in other volumes in this series, but for comparison with the microbiological processes it might be mentioned that pyrolysis and hydrogenation produce mixtures of gases along with oils and carbonaceous char. To obtain full energy recovery all these products must be utilised in some way and the gases can often only be used with special burners. However, a principal drawback to these processes is the high temperatures and pressures involved. These necessitate heavy plant, unsuitable for use on a small scale, on a farm for instance.

For universal use and for the utilisation of as much as possible of the biomass available, methods which can be used on a variety of scales from small to very large are needed, and these methods are provided by biological processes.

BIOLOGICAL METHODS

Biological methods for fuel production are all based on the use of micro-organisms, with or without some additional chemical or physical process.

These methods lead to either a liquid or a gaseous fuel, and the processes being currently experimented on or developed commercially are for the production of the liquid, 'alcohol' (ethanol) or the gases methane and hydrogen.

Production of ethanol by microbiological processes is, of course, as old as mankind, as this is the production of alcoholic drinks. The production of ethanol for use as a petrol substitute is now occurring in Brazil and the substrate used is sucrose. The Brazilian climate is such that sugar cane can be grown on a large scale and production of sugar and subsequently ethanol on a scale commensurate with its use as a petrol substitute is feasible. However, the process is not straightforward, the principal difficulty being that the yeasts which produce ethanol are themselves inhibited by ethanol. Thus the fermentation can produce an aqueous liquid containing only some 10–20% of ethanol. For use as a fuel (or indeed as a high-alcohol-content drink) the ethanol must be concentrated from this aqueous solution by distillation, and distillation is a high-energy-input process. Much effort is now being devoted to increasing the alcohol content in the fermentation and reducing the energy input to distillations or finding means other than distillation for concentrating the alcohol.

In other countries, starchy substrates such as cassava, or cellulosic materials such as domestic waste, are being examined as feedstocks for ethanol production. Here a two-stage process is needed with a preliminary chemical or enzymic hydrolysis of the polysaccharide before the resulting sugars are fermented to ethanol.

Hydrogen is a product of microbial metabolic reactions of various kinds. Hydrogen can be produced by anaerobic fermentation of sugars, as will be seen in the next chapter where its role as an intermediate in methane production is discussed. However, the conversion of the hydrogen contained in sugars to hydrogen gas is never quantitative; part of the hydrogen is contained in other products of the fermentations, various acids or ethanol, and such fermentations are not a really viable method for production of hydrogen as a fuel.

Hydrogen is also a product of photobiological reactions by some bacteria and algae. Although hydrogen production has been demonstrated in small-scale experiments, such systems are many years away from a practical proposition. There is still much work to be done on photobiological hydrogen production and besides practical difficulties there is the fact that experiments are being done in places such as California, an area of intense sunlight. Prospects for photobiological energy production would seem even less promising in a dull, cool country like Britain.

But if a gaseous fuel is to be produced, then methane has advantages over hydrogen. Methane can be burnt in conventional domestic or commercial gas fires or boilers, it can be easily piped and the heat produced by its combustion is, on a volumetric basis, some three times greater than that of hydrogen. And methane is the one gaseous fuel that has been produced microbiologically in large-scale plants for many years. Since this is so, one might ask why is there not more widespread use of microbially-produced methane, why are there not methane plants all around us, and why is the present research and development needed? Some of the reasons will become apparent later in this book, but basically it is a question of cost and the need for fuels.

The large-scale plants in which methane is produced biologically have all been sewage-treatment works. There, the methane-producing plant, the anaerobic digester, is primarily a method of stabilising, and thus reducing pollution from, the sewage sludges produced in the works. The gas produced is a by-product used to heat the digesters and as a source of power.

Although gas from the town septic tank was used to light streets in Exeter a hundred years ago, engineers at the Birmingham, England, sewage works pioneered the use of gas in internal combustion engines for large-scale generation of power. This was developed until in many large sewage works all over the world the digester gas now provides most, or all, of the power required for the works. Generators can supply electricity for lighting and electric motors driving aerators for the aerobic sewage-water treatment, and for sewage pumps. Gas engines coupled to compressors can supply air for sparged-air aerobic-treatment plants. In some cases excess electricity is put into the national or town systems. Some sewage works also ran lorries on the digester gas, but in general as described elsewhere, digester gas is not a convenient fuel for vehicles and so its use fell in face of the more convenient and cheap petrol or diesel oil.

This latter statement is the crux of the story of development of anaerobic digestion as a source of power. Although sewage works continued to use high-rate digestion, only in the big works was it, and of course still is, used as a power source. Small works found it more convenient to use the digester gas only to heat the digesters and to flare-off any unused gas, while using mains electricity or petroleum-based fuels for the works.

Digesters, then, have continued to be used in sewage treatment because they are a relatively simple method of stabilising sewage sludges. They may also provide power for the sewage works, but the cost of the sewage collection and separation and the cost of the digesters themselves are all a part of the necessary sewage system required to enable a modern city to run; the systems are not costed as energy-producing plants.

In other areas and at different times, digestion has been used as a source of fuel when conventional fuels have been unobtainable. During and after the last war digester gas was used as a fuel. Plants were constructed in Germany and elsewhere to generate gas from agricultural and other wastes. But these plants had problems and the return to cheap and plentiful supplies of oil-based fuels in particular made it not economically worthwhile to continue research to overcome these problems and improve their energetic efficiency. Indeed, even if the problems had been solved it is doubtful if any plants could have competed purely on an economic basis with oil at the prevailing low prices, and in matters of convenience, particularly for vehicular use, oil fuels undoubtedly held the lead.

Thus, as sources of power, digesters more or less disappeared. A revival of the digester as a power source began to appear in the late 1950s with the beginning of the Indian Gobar-gas schemes where a hot climate, available man-power and small energy requirements for peasant cooking and lighting, combined with little cheap or available fuel, made very small, non-automated, non-heated digesters a practical possibility. Nevertheless, economics plays a part here and a digester costing only a few pounds is still expensive for a peasant farmer whose income is only a few pounds a year. So research continues, to improve the efficiency of the small digesters and to reduce their cost if possible, and also to develop bigger, communal biogas schemes.

However, what motivated the authors and their colleagues, and some others, to investigate digestion some 12 or 15 years ago was not primarily energy generation but pollution control. It was becoming evident that with the increase in size and in numbers of intensive farming units, pollution from animal excreta, formerly negligible or ignored, could become a serious problem and be controlled by legislation.

It seemed to the authors and their colleagues that digestion, being a process used for reducing pollution from thick domestic sewage sludges, akin in some ways to the farm slurries, could be usefully employed in reducing farm-waste pollution, while the nature of the slurries would make optimum aerobic treatment difficult if not impossible. This expectation has been borne out in practice. Anaerobic digestion has also been tested and used for treatment of polluting factory wastes.

In the pollution control context with farm wastes, the biogas was originally thought of as the by-product which could run the process, thus diminishing its cost, and perhaps provide some surplus energy for use on the farm; but other supplies of energy were plentiful and cheap. The first 'oil crisis' of about 1974 turned attention to alternative sources of energy, and shortages of oil, even if at present mainly politically inspired, and its ever-increasing cost and so the cost of coal and electricity, have kept people's attention on alternative energy sources and have also made these alternative energy sources more desirable and economically more viable.

Whilst anaerobic digestion is now being considered as an alternative energy source, the pollution control aspect must not be forgotten, although it may be difficult to quantify in monetary terms. Pollution control legislation is becoming stricter and is being more rigorously enforced, and this applies to agriculture as well as to industry. So all through this book there will be reference to pollution control aspects of digestion. Not only is this an important part of the digester use; indeed, its original *raison d'etre*, but if some monetary value can be placed on this then it alters the cost of the energy produced by the digester. An aerobic sewage-treatment system has not only capital cost but is costly to run in terms of energy and money. A digester has capital costs, but can be intrinsically cheaper to run in energy terms and it can cover these energy costs from gas produced, and produce a surplus. The large municipal sewage works digester produces energy sufficient for digester heating and running its own pumps, with a surplus for the running of the aerobic treatment systems and other uses. Pollution control in some aspects is a consideration not only of the large intensive farm or industrial digester, it also applies to the small Gobar and similar digesters and is stressed in explaining the advantages of the digester to the farmer.

Agricultural wastes as digester feedstock will, perhaps, seem to be emphasised in this book, but agricultural wastes represent the biggest source of feedstock for energy production by digestion. Modern intensive farming relies on energy inputs in the form of oil, gas and electricity. And unless man is to return to a subsistence feeding level of a monotonous diet, intensive agriculture will have to continue and so the wastes will be present. If these wastes can contribute to the running of the agricultural system or do more than that, then present 'western' standards of feeding can be helped to survive and food improved in underdeveloped areas. Anaerobic digesters can never replace the large power station or produce all a country's static (as opposed to vehicular) energy needs; how far they may contribute to energy needs is discussed in this book and how far they will contribute will depend on the outcome of research and especially developmental work over the next few years.

CHAPTER 3

The Microbiology and Biochemistry of Anaerobic Digestion

The micro-organisms concerned in the production of gas from organic material are bacteria. There is not space here to give a detailed introduction on aspects of bacterial metabolism, it must suffice to remind the reader that there are three main classes of bacteria: those which must have oxygen to grow—the aerobic bacteria; those which can metabolise and grow with or without oxygen—the facultatively anaerobic bacteria; and those which can grow only in the absence of oxygen—the anaerobic bacteria. The bacteria involved in the metabolic pathways leading from waste to biogas are all of the latter class, although as will be seen, all digesters contain aerobic and facultatively anaerobic bacteria as well.

Bacteria can grow in 'closed' and 'open' systems and both are used in biogas production. The closed system is essentially a 'batch' culture. Bacteria are inoculated into a flask or a tank containing suitable nutrients at a suitable temperature and they begin to grow. Their growth is described by the curve in Fig. 3.1. There is a 'lag' phase (a) when there is no, or little growth, a period of active growth (b) and a period (c) when the bacteria stop growing because they have used up the nutrients or have produced substances (say acids) which inhibit further growth. The 'batch' culture has a finite life and must then be started again to continue the life of the bacteria. However, it is possible to continue the active growth phase (b) *ad infinitum* by taking out some of the culture fluid with bacteria and replacing it by fresh nutrients. If this is done continuously the bacteria will grow continuously, and so the culture has no definite life span.

Anaerobic digestion is not a unique 'man-made' bacteriological process; essentially the same microbial activities occur in stagnant pools, in marsh land, in the mud of polluted rivers and in the digestive tracts of herbivorous animals. What the digester engineer and microbiologist tries to do is to increase the rate of these natural reactions and to direct them along one path—the complete conversion of the energy-source substrate into gas.

Whilst the microbial nature of the digestion process has been known for very many years, septic tanks and the later digesters were functioning before the exact nature of the microbes involved, and the chain of reactions

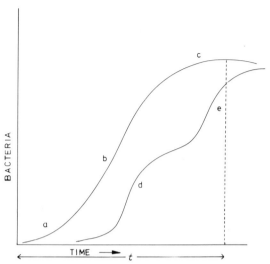

FIG. 3.1. Schematic growth curve of a batch culture; a, b, c are different stages in the time-course of bacterial growth (see text). Curve d–e represents diauxic growth: over d one substrate is being used, and the utilisation of the second substrate is suppressed until the first substrate is used up, when growth starts again on the second substrate, curve e.

they carry out, was investigated. Even now the particular bacteria involved and the biochemical pathways of some reactions are only partly understood.

The microbiology of anaerobic digesters has, both in techniques and determination of the biochemistry involved, depended on the advances made in investigating the microbiology of the rumen system of ruminant herbivores, and this has developed only since the late 1940s. The analogy between the rumen and the anaerobic digester and a detailed description of the micro-organisms involved are discussed in a review by the present authors (Hobson *et al.*, 1974) which also describes the biochemistry of the process in detail and gives references to many original papers. Here, only a brief discussion is attempted and reference is given only to papers making particular points. Further reading will be found in the list at the end of the chapter. As in all scientific work, new knowledge is continually being acquired and some points, at present obscure, may be clarified before this book is published.

Nevertheless, anaerobic digestion remains in many ways an empirical science, depending on practical tests; this is because of the many substrates used, the complexity of these substrates and the reactions involved in their breakdown. In most industrial microbiological processes only one organism is used and this reacts with a medium which, although it may be quite complex in number of constituents, has each constituent defined in concentration and usually defined in chemical constitution. These form the sources of growth energy and the substance of the growing microbial cells, and are all in solution in the original medium, or in the case of oxygen or other gases their rate of solution into the medium can be calculated or determined by experiment. The rates at which these substrates can be utilised are known, as is the extent to which they can be utilised. The growth of the micro-organism can be controlled for optimum production of cells or metabolic product. Mathematical modelling can be applied in design of the plant. By a mixture of laboratory experiments and calculations the design and function of the plant can be precisely controlled.

The substrates for anaerobic digestion are almost entirely wastes containing complex, and generally undefined, materials and the composition of the wastes may vary from day to day, and from situation to situation. Some of the constituents may be defined more by method of analysis than by composition, and the same analytical method can give results that are not comparable from waste to waste. Since many of the substrates are insoluble, physical structure can influence the degree of biodegradability. 'Cellulose' can, for instance, vary from chemically- and physically-treated 'soluble' paper tissues to lignified and virtually non-biodegradable natural straws and straw residues. These kinds of factors make predictions of behaviour of wastes in digesters difficult even if a so-called analysis is available. The only way is to do trial digestions. Since the substrates are complex and variable and the microbial population producing the digestion complex, mathematical modelling of digestion is difficult and rate constants and other factors needed to make the model predictive are difficult to determine.

From the engineering standpoint the digester builder has to cope with inherently intractable sludges which make the design of equipment more a matter of what is physically possible rather than theoretically desirable, and this may only be found out by testing the particular sludge on a full-size piece of equipment. In addition, the plant may have to be tailored to limited costs and operation by few and unskilled personnel.

Nevertheless, the basic design and running of digesters must take into account that they are microbiological production plants, and some knowledge of the bacteria and their reactions and their limitations of growth and factors affecting their growth is necessary. So the microbiology of digestion will be discussed in this section.

THE MICROBIAL POPULATION IN GENERAL

The micro-organisms concerned in anaerobic digestion are bacteria. A few flagellate protozoa have been noted in piggery-waste digesters by the authors, but ciliate protozoa have never been seen. It seems unlikely that the flagellates play any important role in the digestion process. Young & McCarty (1969) stated that ciliate protozoa and amoebae were abundant in their small-scale, experimental anaerobic filters, but the reasons for their presence and what, if any, role they were playing are not known. Since a digester is not run under sterile conditions or with a sterilised feedstock as is a factory fermentation, then all kinds of microbes are continually entering the digester. The mere observation of an organism in digester contents is thus no indication that it is playing any useful role, or even growing, in the digester.

Probably the nearest to a sterilised feedstock which could be used in digesters is the waste-water from fruit, vegetable and meat processing. The water in parts of these processes is used at boiling point and is still at approximately 80° or 90 °C when it runs to waste. This water, if used without cooling as digester feedstock, will be virtually sterile, but if the digester is also fed with cool water from initial vegetable washing or other parts of the plant, then this could convey bacteria to the digester contents.

Macerated vegetable matter, especially crop wastes, will contain many micro-organisms, of the order of a million or more per gram. Many of the bacteria on vegetable matter will be aerobic types commonly found in soil and on vegetation. Aerobic bacteria, requiring oxygen for growth, cannot grow to any extent in digesters. Nevertheless, they will not immediately die, and in a continuous-culture type of digester they will be entering the system continuously. Aerobic culture may give the impression that these bacteria are actually growing and functioning in the digester. The fact that bacteria can exist in a viable, although non-growing or dividing, state for a long time and will then begin to grow and divide when placed in a suitable environment makes assessment of the functional bacterial populations of systems such as digesters very difficult.

Faecal materials contain, when voided, many millions of bacteria from the intestines. These are of a wide variety, and include those anaerobic ones which form the functional digester population. But in addition they contain facultatively anaerobic bacteria which proliferate in the intestines. These can form a large part of the population of digesters. Their influence on a digester population was noted in experiments on piggery-waste digestion. The contents of a domestic sewage digester used as inoculum or 'seed' for the start of a piggery-waste digestion contained *Escherichia coli* (a bacterium found in large numbers in human intestines and faeces) as the predominant facultatively anaerobic bacterium. When the digestion had become adapted to piggery waste as a feedstock the place of the *E. coli* had been taken by facultatively anaerobic streptococci, the predominant facultative bacteria of pig faeces. These streptococci formed 50% or more of the culturable bacteria in the digester contents, but they were present in such large numbers in the feed sludge that this, rather than growth, could have produced the digester population. Indeed, their properties were such that it was difficult to assign a function to them in the digestion (Hobson & Shaw, 1973).

There are, then, many difficulties, apart from technical difficulties, in analysing the bacterial populations of anaerobic digesters. The techniques and theory of microbiological analysis of digester contents have previously been described in detail (Hobson & Summers, 1979) and these involve the co-operation of the methods of biochemistry and bacteriology.

Chemical analysis can determine the nature of the feed to the digester and the end-products of the digestion. Other techniques, such as use of radioactive isotopes and pure compounds as feedstock, can identify some of the intermediates between these feed constituents and the products. By analysis and by analogy with other microbial systems possible reaction pathways producing these intermediates and end-products may be worked out. The bacteriologist determines the properties of the bacteria he isolates and then tries to fit them into the postulated scheme of reactions occurring in the digestion, and the isolation of a particular bacterium may suggest a new reaction or pathway. But generally chemical analysis comes first and the importance of a bacterium is then determined not only by its numbers but whether it can be fitted into the postulated reactions of the system.

All the feedstocks for digesters consist of organic matter of vegetable or animal origin. The bacteria require at least an energy source, a nitrogen source and some salts, and most organic feedstocks being complete animal excreta or vegetable wastes contain all these. If a practically pure carbohydrate, containing little or no nitrogen compounds, say paper pulp, was used as feedstock, then nitrogen compounds would have to be added to it. So in general the constituents of digester feedstocks are taken to be those shown in Fig. 3.2, and the reactions taking place as those depicted also in the figure.

Nearly all the biochemical and microbiological work on digestion has been done on sewage or farm-animal-waste digesters, or in a few cases, small-scale digestions using artificial mixtures of the feed constituents

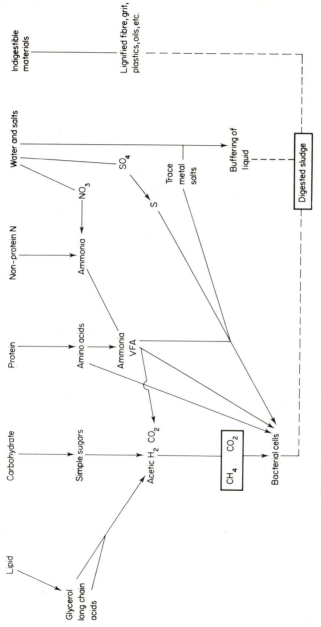

FIG. 3.2.   The main reactions occurring in an anaerobic digestion.

shown in Fig. 3.2. So the data in the next paragraphs are taken from such work. There is no doubt, however, that reactions in all digesters are similar and the important bacteria are of the same or similar types.

THE BREAKDOWN OF CARBOHYDRATES

Wastes from slaughter-houses and meat-processing factories might be thought to contain little carbohydrate, but animal tissues can contain some, and the wastes in general also include all the residues from the animal slaughtering and carcass dressing—gut contents and secretions, and the actual excreta and bedding from animal-holding pens. These latter wastes all contain carbohydrates in much the same amount as the 'pure' animal excreta used in digesters. The composition of piggery waste is shown in Table 3.1.

Thus, carbohydrate plays a major part in digester reactions and its breakdown, outlined in Fig. 3.2, will be followed in detail as this breakdown can be a main rate-controlling process in digestion.

The carbohydrates of all digester feedstocks are principally polymers, insoluble or of limited solubility in water. Large amounts of soluble carbohydrates such as ordinary sugar (sucrose) or the residues from sugar refining, molasses, are probably best used as substrates for rapid fermentation to alcohol, or for single-cell protein (SCP) production. Wastes (sulphite liquors) from wood-pulp production for paper making contain soluble sugars, but the principal microbiological use for these has so far been in SCP production, although they could be a digester feedstock as seen later. Wastes from fruit and vegetable processing will contain sugars, but these also contain the insoluble carbohydrates of peels and pods and other plant structures.

The wastes that seem to offer the greatest possibilities for digestion and biogas production are animal excreta, domestic sewage and garbage, and crop residues or crops grown for digestion—the carbohydrate of all these is essentially polymeric.

Sewage sludges used in digesters contain as well as the primary sludge (essentially larger particles of faeces, toilet papers and other debris), sludge from the aerobic treatment systems which consists of large amounts of bacterial cells as well as undegraded faecal and other particles from the sewage. These aerobic bacteria are slimy and the slime consists in part of complex polymeric carbohydrates. Excreta of animals and humans contain bacteria and intestinal secretions and cells sloughed-off from the intestinal

TABLE 3.1

SOME REPRESENTATIVE ANALYSES OF SEWAGE SLUDGES AND
PIGGERY WASTES (TWO SAMPLES)

Sewage sludge solids (%)[a]

Lipid (ether soluble)	34·40	Lipid	25·2
Water soluble	9·52	Crude fibre	10·8
Waxes, resins etc.	2·49	Cellulose	1·4
Hemicellulose	3·20	Humic acid	4·0
Cellulose	3·78	Ash	39·2
Lignin	5·78		
Crude protein	27·12		
Ash	24·13		

Piggery wastes[b]
Solids (%)   *Solids* (%)

Lipid	13·72	Lipid	7·7
Cellulose	15·48	Crude protein	20·9
Hemicellulose	22·31	Cellulose	22·9
Lignin	8·12	Hemicellulose	20·8
Crude protein[c]	19·88	Lignin	10·1
Ammonia N	2·15	Ash	17·6
P	1·67		
K	0·93		
Ash	13·96		
Starch	0·00		

Supernatant (% wt/vol)

Total N	0·25
Ammonia N	0·21
P	0·014
K	0·06

From figures given by [a] Pohland (1962), Buswell & Neave
(1930); [b] Summers & Bousfield (1980), Hobson *et al.*
(1974). [c] Non-ammonia N × 6·25.

walls. Bacteria can contain starch-like storage carbohydrates and sugar
molecules of different kinds incorporated in components of the cell. Gut
secretions contain complex carbohydrates, the mucopolysaccharides.
Some of these carbohydrates may be virtually unutilised in the digestion
process; others may be liberated as the faecal bacterial cells and intestinal
cells die and disintegrate. However, the most important source of
carbohydrate in faecal materials, especially animal excreta, is the residue of
vegetable matter from the feedstuffs passing through the gut, although
toilet and other papers and articles made from cotton will also contribute
some carbohydrate to the domestic sewage sludge.

Human beings have only a limited ability to utilise polymeric carbohydrates and starch alone can be degraded by intestinal secretions. Starch, particularly as this is cooked, is almost completely utilised in its passage through the digestive tract. On the other hand, the fibre of breads, breakfast cereals, etc., and of vegetables and fruits is almost completely unattacked in the digestive tract and so is voided in the faeces. Farm animals, because they contain organs where a slow microbial fermentation of the feed takes place, can digest plant fibres including grain husks. However, the digestion of fibre is never complete and the residues of fibres left in the faeces are those most resistant to bacterial attack, so the vegetable carbohydrates in animal faeces are a poorer substrate for the bacteria of an anaerobic digester than are those of human faeces. Starch in animal feeds is more or less completely removed in the digestive tract and little reaches the faeces.

Macerated fresh vegetable matter which could be used by itself or mixed with animal excreta will contain some soluble sugars, but the amount varies with stage of growth of the plant. Grasses, for instance contain some 6–7 % or sometimes more of the dry matter as soluble sugars (Waite & Gorrod, 1959). Sugar-beet or sugar-cane, of course, contain large amounts of soluble sugar, but this is more likely to be used as a feedstock for other fermentation processes and when the sugar is removed only fibrous structural tissue remains. Some other root vegetables contain soluble sugars; turnips for instance contain about 35 % of the dry matter and the stems of plants such as kale also contain about 25 % soluble sugars in the dry matter. These types of vegetable are generally used as feeds for farm animals and are likely to be used as digester feedstock only on a small scale to use up rotten or undersized crops.

Of more likely use as digester feedstocks are the starchy tuberous plants, potatoes, cassava and so on, even though potatoes are only likely to be used when undersized or otherwise unsuitable for human consumption.

Thus, with the possible exception of starch from tubers, it would seem that vegetable matter, whether grown especially as an 'energy crop' in tropical or subtropical areas or as residues from crops grown for human or animal food, is likely to be leaf and stem material. This consists mainly of polymeric carbohydrates or 'polysaccharides' of the plant structure; cellulose, hemicelluloses and pectins. Seaweeds are giant algae and, like the microscopic algae, also a possible feedstock. They have chitin as a structural polysaccharide and contain the polysaccharide alginic acid.

These polysaccharides are composed of simple sugars linked together to form chains many hundreds or thousands of sugar units long. The chains may be so-called 'straight' chains, that is sugar molecules joined 'end-to-end' to form one long polymer, or they may have sugars joined 'end-to-end' with other chains of sugars joined on to the sides of some of the sugars of the main chains, thus forming a complex, three-dimensional array of chains. In addition, the straight-chain molecules may be arranged side-by-side to form bundles of chains, as in cellulose.

The polymeric carbohydrates vary, thus, in physical structure and in solubility, and they also vary in the simple sugars which make up the polymer. The digester bacteria can use only simple sugars, either mono- or disaccharides, as sources of energy, so the polymers must be degraded by hydrolytic action of enzymes from the bacteria into the simple sugars before being absorbed and utilised by the bacteria. All the sugars of which plant polysaccharides are composed can be utilised by bacteria so that the monomer or monomers making up the polysaccharides are not of vital importance in digestion. Some bacteria can utilise many different sugars, and so could be found in a digester with any feedstock, merely utilising whichever sugar was present. Some bacteria are more specialised and so these bacteria would be expected to increase or decrease in number according to the sugar present in the feedstock. But whichever way it is done a digester population which can use the sugar presented will develop.

On the other hand, the monomers and the linkages between them, and the structural form of the polysaccharides are of great importance in the primary step in digestion, the hydrolysis of the polysaccharides. The assimilation and metabolism of mono- and disaccharides from solution is a relatively rapid process with bacteria. These sugars, e.g. glucose, can be assimilated at a rate which enables bacteria to double in size and divide in 20 or 30 minutes. However, if the glucose is in a polymer the rate of hydrolysis of the polymer to give glucose may be so slow that the glucose becomes available to the bacteria only at a rate which enables them to divide once every 24 or 48 hours. The rate-limiting step in carbohydrate fermentation in digesters is then often the hydrolysis of polysaccharides.

The monomers of the polysaccharides and the way they are linked together become important in this hydrolytic step because different enzymes are required according to the monomer and the linkage, and these latter also determine the physical structure of the polysaccharide and its solubility.

While hundreds of different kinds of bacteria will use glucose as an energy source, less can use the glucose polymer starch, and even fewer can use the glucose polymer cellulose. The number of bacteria which can hydrolyse and utilise the polysaccharides composed of other sugar residues is also small. However, the hydrolysis of polysaccharides occurs outside the bacterial cell and the resulting sugars diffuse into the culture, or digester, liquid where they can be absorbed and utilised both by the hydrolytic bacteria and by other surrounding bacteria which have no polysaccharide-hydrolysing ability.

Thus, in a digester the rate of hydrolysis of a polysaccharide which constitutes a main proportion of the carbohydrate of the feedstock can influence the rate of growth and metabolism of a large part of the digester flora. If carbohydrate is the main source of fermentation acids and hydrogen, then the rate of growth of the methanogenic bacteria can in turn be governed by the rate of metabolism of the carbohydrate.

The degradation of cellulose which is a main constituent of plant structural polysaccharides can be used as an illustration of these points. While starch is a polymer made up of glucose units joined by $\alpha1:4$ links and can be dissolved in water, cellulose has $\beta1:4$-linked glucose units and is insoluble in water. Some of the $\alpha1:4$-linked glucose chains in starch are relatively short and are linked together by $\alpha1:6$ links to form an irregular, branched structure. Other $\alpha1:4$-linked chains are much longer but are coiled. This, overall, tends to give starch granules a more 'open' structure than that formed by the long, straight, β-linked chains of the cellulose molecules which can lie close together in bundles which make up fibres. The open structure of starch is more easily dispersed into solution in water than is the close-packed structure of the cellulose fibre and is more accessible to bacteria and their enzymes even when not completely dissolved. Starch is not too easily soluble in water but can be made easily soluble even in cold water by making the polymer molecules shorter by a partial hydrolysis in, say, dilute acid. This breaks some of the links between the' glucose molecules. Cellulose also can be more easily dispersed in water, if not rendered actually soluble, by acid hydrolysis. If the molecules are in solution then they are free to move in the water and can readily come into collision with enzyme molecules also in solution which can attach to the carbohydrate molecules and disrupt the glucose linkages. The acid treatment can, by breaking interglucose linkages, break up the regular structure of the mass of molecules making up a cellulose fibre and allow the enzyme molecules more ready access to the glucose chains. The treatment also forms more chain ends from which the enzyme molecules can begin their hydrolysis.

The bacteria which can hydrolyse cellulose vary in the amount of cellulose-hydrolysing enzyme (cellulase) which they secrete and thus in their rate of attack on any particular cellulose. The amount of cellulase which bacteria can secrete will also affect their ability to hydrolyse different forms of cellulose and it also seems that 'cellulase' consists of more than one enzyme and that at least two cellulolytic enzymes are required for the degradation of some of the forms of cellulose which occur in native plant structures. Bacteria appear to vary in the relative amounts of the enzymes which they produce and thus vary in their ability to attack different forms of cellulose. In the rumen the bacteria classed as 'cellulolytic' from their ability to attach to, and grow on, the prepared cellulose used in laboratory media, consist of types varying in the amount and the constituent enzymes of cellulase that they secrete, and a similar situation probably exists in digesters, although this latter has not been so extensively studied. Electron micrographs show that many different kinds of bacteria are present in, on and around plant fibres being degraded in the rumen and it seems likely that these bacteria are 'collaborative' in that not only will non-hydrolytic bacteria be living on the products of the hydrolytic bacteria, as explained before, but also bacteria with different hydrolytic properties will attack different cellulosic structures, or, some will attack cellulose structures initially partially degraded or 'opened-up' by those with a more complete complement of the mixture of cellulolytic enzymes.

However, although the digester bacteria will consist of a mixture of different types, capable as a whole of degrading all forms of cellulose, the absolute rate at which the cellulose substrate is attacked will still depend on its physical form.

In domestic sewage, cellulosic materials will consist in part of food residues in faeces. These residues will be of vegetable matter which has been cooked in some way, thus the cellulosic fibres will have been physically damaged by the pre-cooking chopping and grinding and further disorganised and possibly partially hydrolysed by the boiling or other heating they have been exposed to. The ease of bacterial hydrolysis has thus been increased. Other forms of cellulose will be toilet papers, which are purified wood-cellulose, chemically and physically treated to make it 'soft' and 'soluble'. Thus, overall, cellulose of domestic sewage has been rendered relatively easily degradable by bacterial enzymes.

On the other hand, cattle are fed mainly on untreated cellulosic materials, grasses and other plant leaves and stems, and straw. Living plant structures must be relatively resistant to microbial attack otherwise the plants could not exist, and even when they are dead, this resistance persists. The resistance to attack is not only conferred by the orderly and close arrangement of the cellulose and hemicellulose molecular structures, but by the presence of substances inherently resistant to microbial enzymes; waxes, lignin and even inorganic materials such as silica. Lignin is a very complex compound, whose structure is still not fully determined and it is, to all intents and purposes, unattackable by anaerobic bacteria. The lignin which is intimately connected with the cellulosic structures increases in amount as plants grow older and acts as a barrier to bacterial attack on the cellulose. It seems likely that it not only impedes the passage of cellulolytic enzymes but prevents the cellulolytic bacteria adhering to the plant fibres, (see Hobson, 1979 for review) and adherence is a prerequisite for optimum bacterial attack.

Lignification prevents microbial degradation of cellulosic fibres in the rumen, and the more lignified the plant material is (e.g. the older a grass) the less it will be degraded. The plant-fibre residues finally excreted in the faeces will thus be the lignified material most resistant to bacterial attack. Straw is a very hard, lignified residue left after growth of the cereal plant has ceased, and it is very resistant to microbial attack, as demonstrated by its use in roof-thatching and other materials. Even in the rumen of cattle less than half of it can be digested and what is left in the faeces is very resistant to further microbial attack. Pigs are now not usually fed on fibrous material but on grains, and so grain husks, of lignified cellulosic material, are the residues left in the faeces. If bedding is used for farm animals it is usually of straw, wood sawdust or shavings. This will be mixed with the excreta as it is collected from the animal houses.

Farm animal excreta thus contains cellulosic materials inherently more difficult to degrade than those of human excreta. Ghosh & Klass (1978) in studying two-stage digesters (see later) found that municipal sewage sludge (90 % activated sludge and 10 % primary sludge) could be hydrolysed and fermented to acids in a detention time of 0·5 to 1·0 day. The rate of fermentation of the sludge was about twice that of a purified cellulose and a detention time of about 1·7 days was suggested for the cellulose fermentation. (The conversion of the acids to methane required a detention time of 6·5 days.) The cellulose they were using was virtually 100 % digestible. However, in piggery-waste digestions the cellulose in the waste was only 41 % digested at a detention time of 10 days (Hobson, 1976). Barley straw (added to piggery waste) was only 35 % fermented at a detention time of 20 days and was much less degraded at a 10-day detention time (Summers, 1978; Hobson, 1979). These digestions, like those of Ghosh & Klass were mesophilic ones run at 35 °C. Pfeffer (1978) reported that although degradation of wheat straw in a thermophilic digestion at 59 °C increased with increasing detention time, the straw was about 38 % degraded in a detention time of 13·7 days and he calculated that at infinite detention time the straw would be only 50 % degraded.

In all these cases the rate of conversion of the fermentation acids to methane would be essentially the same, so the results show the differences in overall rates of digestion which can be caused by differences in the biodegradability of cellulosic materials. Conversion of acids to methane would be rate-limiting in the digestion of sewage and purified cellulose while degradation of the cellulosic substrate would be rate-limiting in a straw digestion.

The influence of lignin on cellulose degradation and gas production in digestion is also shown by the results of Robbins *et al.* (1979). They tested the digestion of cattle waste and cellulosic substrates. Their laboratory-scale digesters were run at a 16-day detention time and 37 °C. Five percent Total Solids (TS) cattle waste was mixed with 1 % straw (i.e. a total of 6 % TS). The waste plus straw gave a similar gas production to 6 % TS cattle waste alone, 39 % higher than the 5 % TS cattle waste. However, 1 % of straw which had been delignified with acid chlorite gave a 94 % higher gas production than the 5 % cattle waste alone, and 1 % of purified cellulose (filter paper) gave the same gas production as the delignified straw. The calculated digestion of organic matter in the cattle waste or the waste plus straw was about 32 %, but with the delignified straw or cellulose the organic matter digested was 44–45 %. The delignified straw, or prepared cellulose, was almost entirely digestible while the untreated straw was only about 32 % digestible.

Lignin was also shown to be almost entirely undegraded in anaerobic digesters. The non-degradability of lignin is determined by its complex structure, not by the non-degradability of the different kinds of molecules which are linked together in the polymer. Healy & Young (1979) found that a number of compounds of the types which make up the lignin polymer were themselves degraded to methane and carbon dioxide by the mixture of bacteria from a domestic sewage-sludge digester.

The extents of hydrolysis of celluloses quoted above were determined by chemical analyses of the feedstocks and the digested sludges or by calculations from the gas production, and bacterial counts can be used to support these observations.

Hungate (1950) found no cellulolytic bacteria in domestic sewage sludge but found 0·8 to 2 × 10³ cellulolytic bacteria per ml of digesting sludge. Maki (1954) also found no cellulolytic bacteria in raw domestic sludge but 1·6 × 10⁴ to 9·7 × 10⁵/ml in sewage digesters. Hobson & Shaw (1974) found no cellulolytic bacteria in piggery waste, but found 4 × 10⁴ to 4 × 10⁵ cellulolytic bacteria per ml of sludge from a mesophilic piggery-waste digester. In sludge from a domestic sewage digester they found 4 × 10³ cellulolytic bacteria. This sludge was used to start a piggery-waste digester and as the bacterial population adapted to digestion of the piggery waste of higher cellulose content than domestic sewage sludge (Table 3.1) the numbers of cellulolytic bacteria increased to the 10^5/ml level. There were eleven different types of bacteria in the cellulolytic bacterial population of the piggery-waste digester. The variation between the bacterial counts reported for the domestic sewage digesters could be due to variation in the cellulose content of the sewage sludges (for examples see Table 3.1). However, as will be discussed later, cycles of bacterial activity take place in systems such as digesters and the numbers of bacteria at particular times can be affected by the way in which the digester is loaded as well as by the overall detention time and by the loading rate. The cyclic variation in activity is the reason why counts, such as those above, of 4×10^4 to 4×10^5 are given; these are maximum and minimum figures from a series of counts done over a period of some weeks on a digester running in a 'steady-state' (see later).

It was previously mentioned that cellulose in plant structures is accompanied by hemicelluloses, and hemicellulose-degrading bacteria were isolated from piggery-waste digesters in numbers similar to those of the cellulolytic bacteria (Hobson & Shaw, 1971, 1974), but they appeared to be predominantly of one species. This could be a reflection of the fact that, so far as is known, hemicellulose does not require the complex enzyme system of the cellulases for its hydrolysis.

It was also previously stated that there were more bacteria capable of attacking starch than those capable of attacking cellulose, and this is reflected in studies on anaerobic digesters. Amylolytic bacteria were found in numbers of 10^4–10^5/ml in piggery-waste digesters, and in numbers of about 10^8/ml in the piggery-waste feedstock of the digesters (Hobson & Shaw, 1971, 1974). There was no starch in the digester feedstock and thus none in the digesters, so the high numbers of starch-degrading bacteria probably reflect the fact that starch hydrolysis is a relatively common property of bacteria. It is possible, however, that growth of amylolytic bacteria was encouraged by starch passing through the pig intestines and these bacteria were voided in large numbers in the faeces, and there could have been growth of bacteria on residual starch (with destruction of this starch) in the faecal slurry while it was standing under the pig houses. Torien (1967) also found starch-degrading bacteria in domestic sewage digesters while analyses show an absence or virtual absence of starch in domestic sewage sludge (Table 3.1), supporting the first suggestion, that starch hydrolysis is a fairly common property of bacteria.

Bacteria fermenting glycerol formed by hydrolysis of lipids were also found in piggery-waste digesters (Hobson *et al.*, 1974).

Thus the limited bacterial analyses of digesters done so far have shown that bacteria which can degrade the main carbohydrates of plant structures can be found in anaerobic digesters. The bacteria found in the experiments quoted above could ferment carbohydrates other than those specifically mentioned, including polymeric carbohydrates, and there is no doubt that detailed studies on digesters with feedstocks containing the more unusual carbohydrates such as the alginates of seaweeds would show that the digester floras contained bacteria capable of hydrolysing these carbohydrates.

The polysaccharide-fermenting bacteria also ferment various sugars, not all structurally related to the polysaccharides, and tests using media containing only simple sugars show that the digesting sludge contains many more sugar-fermenting than polysaccharide-fermenting bacteria. Counts of about $1-7 \times 10^8$ bacteria/ml of domestic-digester sludge have been found (Mah & Susman, 1968; Kirsch, 1969), while a piggery-waste digester contained between 6.4×10^5 and 8.4×10^6 bacteria/ml (Hobson & Shaw, 1971, 1974). These bacteria would be living primarily on the sugars formed by polysaccharide breakdown.

Each type of bacteria when grown by itself forms a number of fermentation products, and if these were also formed in the mixed population of the digester one might expect the products of the first hydrolytic and fermentative stage of digestion to be a mixture of hydrogen, carbon dioxide, formic, acetic, propionic, butyric, lactic and succinic acids and ethanol. In actual fact the acids found in a digester are predominantly acetic, with little or no propionic or butyric acids and no lactic or succinic. Methane and carbon dioxide are the gases, with, possibly, traces of hydrogen. Hydrogen is rapidly converted to methane, so would not be expected to accumulate and the presence of acetic rather than other acids could be accounted for primarily by two mechanisms.

Lactic and succinic acid are fermented by some bacteria to acetic and propionic acids. For instance 3×10^7 bacteria fermenting lactic acid to acetic and propionic acid were found per ml of piggery-waste digester sludge (the digester was loaded at a higher rate than that used for the counts of cellulolytic and other bacteria quoted from Hobson & Shaw, 1974) by Hobson *et al.* (1974). Thus any lactic or succinic acid formed by fermentation of sugars would be immediately used up. Ethanol can also be converted eventually to methane and carbon dioxide. However, the formation of lactic and succinic acid as well as ethanol and propionic and butyric acids will tend to be prevented by the action of the methanogenic bacteria in utilising hydrogen.

The primary breakdown of sugars in fermentations is to pyruvic acid, with liberation of hydrogen in the form of a hydrogen-carrier complex. This hydrogen could then be used to reduce pyruvic acid to propionic acid. Pyruvic acid can also be reduced to ethanol by a different pathway:

$$C_3H_4O_3 + [2H] \rightarrow C_2H_6O + CO_2$$

or to lactic acid:

$$C_3H_4O_3 + [2H] \rightarrow C_3H_6O_3$$

Pyruvic acid can also be converted to butyric acid (via acetic acid derivatives):

$$2C_3H_4O_3 \rightarrow C_4H_8O_2 + 2CO_2$$

or converted to succinic acid (via propionic acid):

$$C_3H_4O_3 + CO_2 + 2[2H] \rightarrow C_4H_6O_4 + H_2O$$

The production of acetic acid from pyruvic acid is:

$$C_3H_4O_3 + H_2O \rightarrow C_2H_4O_2 + H_2 + CO_2$$

The hydrogen can then be used by the methanogenic bacteria to form methane and water:

$$4H_2 + CO_2 \rightarrow CH_4 + 2H_2O$$

If the methanogenic bacteria are growing with the sugar-fermenting bacteria the removal of hydrogen will induce the bacteria to form more hydrogen, thus instead of a mixture of acetic and propionic acids:

$$C_6H_{12}O_6 \rightarrow C_2H_4O_2 + C_3H_6O_2 + CO_2 + H_2$$

acetic acid would be produced:

$$C_6H_{12}O_6 + 2H_2O \rightarrow 2C_2H_4O_2 + 2CO_2 + 4H_2$$

The hydrogens formed in the initial split of glucose to pyruvic acid would be released as hydrogen gas and more hydrogen would be released in the formation of acetic acid. The $4H_2$ would then be combined with CO_2 to form methane. In a similar way the production of ethanol, lactic acid and the other reactions shown above, would be displaced in favour of acetic acid and hydrogen production.

The equations above show a complete 'pulling' of the glucose breakdown to form acetic acid, hydrogen and carbon dioxide. In practice, in the mixed bacterial population of the digester the fermentations would not be completely biased towards acetic acid and hydrogen, but sufficient experiments have been done in laboratory cultures with mixtures of methanogenic bacteria and bacteria which ferment sugars to the various acid products to show that this 'pulling' of the fermentation by removal of hydrogen is very likely to occur in digesters (e.g. Iannotti *et al.*, 1973; Chung, 1972; Latham & Wolin, 1977).

The opposite reaction can also occur. If the methanogenic bacteria cease growth for some reason or another, then hydrogen could begin to accumulate. This would drive the fermentation away from acetate production and towards the production of more propionate (Henderson, 1980).

The demonstrations of the effects of abstraction or introduction of hydrogen in the mixed-acid fermentations have been done with rumen bacteria and these are essentially similar to, if not in all cases the same species as, digester bacteria. Analysis of the acids and gases in digesters also suggests that the reactions occur.

In a batch digestion of undiluted piggery slurry which finally became acid and stopped, the gas composition after six days of digestion, when the production rate was just beyond maximum, was $81\% \, CO_2$, $10\% \, CH_4$ and $9\% \, H_2$. After 14 days when gas production was almost zero it was $95\% \, CO_2$, $4\% \, H_2$ with only traces of CH_4 and other gases. Acid and gas productions were both high during the first five days of this fermentation, but the methanogenic bacteria to convert the acids to methane were not present in sufficient number to use the acids as they were produced and so these accumulated and the digestion became too acidic for further bacterial growth (Shaw, 1971; Hobson & Shaw, 1974). When a more dilute piggery slurry was used to start a digestion from water by setting up a continuous-culture system in which small amounts of the waste were added daily and an equal volume of the digester contents removed, the gas produced in the early stages, before any appreciable numbers of methanogenic bacteria could be detected, contained up to 25% of hydrogen. When the digestion was fully developed hydrogen was present only in trace amounts and the gas was almost entirely methane plus carbon dioxide (Hobson & Shaw, 1971).

These results show that hydrogen is a product of the first stage of the digester reactions, the fermentation of substrates in the waste. In the case of the piggery wastes the fermentable substrates are nearly all carbohydrates, and little lipid is present (see later).

In the piggery-waste digesters being developed from water, acetic, propionic and butyric acids were present in the early stages in proportions of about 10:3:2. As the digestion developed and methane was produced in quantity, the acid concentration dropped with acetic becoming the main acid with no butyric and only traces, if any, of propionic (Shaw, 1971). On the other hand, in a piggery-waste digester being changed to a very low solids waste, gas production fell off exponentially suggesting a washout of the methanogenic bacteria. On continued loading, gas production remained very small or zero and acids built up. Measurements at one point during this time when the acid concentration was 2100 mg/litre, showed that 94·3 % of this was propionic acid, 4·3 % was acetic and 1·4 % butyric. Change to a waste of higher solids content gradually restored gas production and the acid concentrations returned to its normal value of about 300 mg/litre, nearly all of which was acetic (Hobson *et al.*, 1974).

In experiments on two-stage digestion, where the fermentation stage is separated from the methanogenic stage in two digesters (see later), Cohen *et al.* (1979) used a synthetic waste containing glucose as the only fermentable substrate, and inocula of bacteria from a municipal sewage works. The products of the fermentation taking place without the methanogenic bacteria were hydrogen and carbon dioxide, with butyric as the principal acid, followed by acetate (in proportions 57·2:40·5) with formic acid (8·2), ethanol (2·6), propionate (2·1) and lactic acid (0·4).

Thiel (1969), using small-scale amounts of bacteria from an anaerobic digester showed that when small amounts of chloroform, carbon tetrachloride or methylene chloride were added, methane production was inhibited and hydrogen was produced.

These results all support the suggestion that in the absence of methane production the initial fermentation of carbohydrates is a mixed-acid, hydrogen and carbon dioxide production.

THE BREAKDOWN OF NITROGENOUS COMPOUNDS

The hydrolysis and fermentation of carbohydrate just described provides growth energy for many of the digester bacteria. Although municipal sewage sludges tend to be high in fats (Table 3.1), in the materials being considered as the main subjects for biogas energy production the fat content is low and by far the principal substrates for fermentation are carbohydrates. But to grow on the energy provided by this fermentation the bacteria need nitrogen to build the proteins, nucleic acids and nitrogenous constituents of cell walls, which form the cell.

Bacteria in general are about 90 % water, as are most organisms. Of the dry weight of material about 10 % is nitrogen, 50 % carbon, 10 % hydrogen and 20 % oxygen. The cell constituents also contain about 10 % inorganic matter in the form of metallic salts; sodium, potassium, calcium, etc., chlorides, sulphates and phosphates. They also have small amounts of metals such as zinc and magnesium and iron as constituents of enzymes and various cell structures, and sulphur as a constituent of some amino acids.

The cell carbon is part of all the molecules making up the cell. It is obtained in a number of ways by the different types of bacteria. Some comes from the carbohydrate of the fermentation substrates. With some types of anaerobic bacteria part (up to possibly about 20 % in some cases, e.g. Hobson & Summers, 1967) of the substrate carbohydrate disappearing does not appear as fermentation products but is assimilated and converted into cell constituents. But in many cases the substrate carbohydrate is virtually 100 % converted into fermentation products and used as an energy source. Such bacteria obtain cell carbon from other materials in the culture medium. Amino acids are often a principal source, carbon dioxide can also be used, while some bacteria use acetic acid or the higher volatile fatty acids. For instance, some of the methanogenic bacteria can form up to about 60 % of their cell carbon from acetic acid (Bryant *et al.*, 1971).

But whatever the source of cell carbon, the nitrogen comes from two sources, either amino acids or ammonia. Some bacteria must have amino acids, some can use amino acids and ammonia, some can use either one or the other depending on which is available, and some can use only ammonia. These latter are the ones that have to obtain cell carbon from medium constituents other than amino acids. In anaerobic digesters ammonia is probably a major source of cell nitrogen. Shaw (1971) found that the majority of his bacterial isolates from piggery-waste digesters could use ammonia as nitrogen source.

The nitrogen in the digester feedstock consists of what can be for convenience described as 'protein' and 'non-protein' nitrogenous compounds (Fig. 3.2). This material is, however, complex in most cases. In feedstocks containing human or animal excreta the nitrogen-containing materials include intestinal bacteria, gut secretions, sloughed-off intestinal cells in the faeces, and constituents of urine. The faecal material will contain true proteins, and also nucleic acids, cell walls and other parts of cells. These latter besides containing amino acids, purines and pyrimidines and amino sugars, all of which have nitrogen as a constituent, also contain carbohydrates, phosphate and lipids linked to the nitrogenous compounds or existing as part of a complex multi-layered or other structure. As the cells die, enzymes in the cells themselves can degrade these complex materials and liberate some of the nitrogenous constituents of the complexes which can then be utilised by bacteria either as such or when further degraded. Intestinal secretions contain complexes of carbohydrates and proteins which can possibly be used at least to some extent by the digester bacteria. In domestic sewage sludge the bacteria from the aerobic sludge will be an addition to the bacteria of the faeces. Urines from different animals and humans vary to some extent in composition but will contain, depending on the source, urea, creatinine, uric acid and ammonia; all these are non-protein nitrogen compounds. Faeces contain ammonia.

Faecal material will also contain some proteins from feedstuffs which have escaped digestion in the gut. These can be of vegetable, animal or fish origin. Again, these will be accompanied by some non-protein nitrogenous compounds.

Fresh vegetable matter used as all or part of a digester feedstock will contain true protein, amino acids complexed with polysaccharides and purines and pyrimidines.

The amount and nature of nitrogenous constituents of excreta can change as it is stored. So the composition of slurries fed to anaerobic digesters may not be the same as that of freshly-voided excreta. In particular, bacterial action in collecting troughs and tanks may result in degradation of proteins to amino acids and then to ammonia, while urea and other non-protein nitrogen compounds can be rapidly degraded to ammonia. Ammonia may then be lost by volatilisation. The extent of such changes will depend on the time for which the wastes are stored, but some changes can take place in a few hours, especially in warm climatic conditions.

In the digester, proteins are initially degraded to either separate amino acids or groups of two or three amino acids; these latter may be taken up and utilised by some bacteria, or be further hydrolysed to their constituent amino acids. Such proteolysis has been demonstrated chemically, and confirmed bacteriologically.

Harkness (1966) found proteolytic, Gram-positive bacteria in numbers of about 7×10^4/ml in digesting sludge and Torien (1967) found proteolytic activity to be associated with Gram-positive sporing rods, although some other proteolytic types were also isolated. Kotzé *et al.* (1968) found that proteolytic enzyme activity varied with the type of waste being digested, but this was probably a reflection of the substrate levels and total bacterial activity rather than a specific change in numbers of proteolytic bacteria. Siebert & Torien (1969) isolated proteolytic bacteria in numbers of 6.5×10^7/ml from a domestic sewage digester. Again the principal types were Gram-positive rods (Clostridia), but other Gram-positive rods and cocci were also isolated. The principal proteolytic bacteria isolated from digesting piggery waste were also Clostridia and these were found in numbers of about 4×10^4/ml (Shaw, 1971; Hobson & Shaw, 1971, 1974). The results so far obtained seem to suggest that the main proteolytic bacteria in digesters are Gram-positive whereas those of the rumen are Gram-negative. On the other hand the principal carbohydrate-fermenting bacteria appear to be Gram-negative as they are in the rumen.

The bacteria isolated as the only type fermenting hemicellulose in piggery-waste digesters were identified with the rumen species *Bacteroides ruminicola* and this bacterium could play a major role in the degradation of amino acids to give ammonia, as it does in the rumen.

Bacteria concerned in the pathways of breakdown of proteins have thus been isolated from digesters, but the studies have not been complete enough to demonstrate bacteria concerned in the metabolism of the minor non-protein nitrogen compounds. However, by analogy with the rumen one might expect that bacteria hydrolysing urea to ammonia and breaking down purines and pyrimidines to acids, carbon dioxide and ammonia would be found in digesters. Anaerobic bacteria which carry out these reactions and which could live in digesters are known (Hobson *et al.*, 1974). Nitrate could also be reduced to nitrite and then to ammonia, in digesters, just as sulphate can be reduced to sulphide, and sulphate-reducing bacteria have been isolated from digesters (Torien *et al.*, 1968). Sulphide, from sulphate or sulphur amino acids will act as a reducing agent, keeping the Eh of the digesting sludge low.

THE BREAKDOWN OF FATS

Fats (lipids) are found in most digester feedstocks as they are constituents not only of animal matter but also of vegetable material. The large amount of fat found in domestic sewage sludge (Table 3.1) is a reflection of the meat-eating habits of humans and their high consumption of other fatty materials such as butter. Fats are digested in the human intestines but some will remain unattacked to be voided in the faeces. In general the feeds of farm animals contain only the lipids of vegetable matter, although some animal fats will be contained in feed protein supplements such as meat and fish meals. Again fats are digested by the animal but some will escape digestion to appear in the faeces. Slaughter-house and meat-processing-factory wastes would, of course, be expected to contain large amounts of fat.

But besides residues from food lipids, faecal wastes will also contain the lipids of the intestinal bacteria, and these can amount to some 5–10 % of the bacterial weight. Like the bacterial polysaccharides and nitrogenous compounds these lipids will be released for attack by the digester bacteria as the intestinal bacteria die and disintegrate in the faeces before and after it is voided.

However, although lipids are degraded in digesters there will always be some in the digester output, that which is contained in the digester bacteria. If the digester feedstock is faecal material or aerobic sludge then the mass of bacteria coming into the digester is probably much the same as that going out, so that although some bacterial lipids will be recycled, any change in lipid content will be due to degradation of non-bacterial lipids in the feedstock. If the feedstock contains few bacteria then the digester bacteria will form some lipids from carbohydrates and other non-lipid substrates in the feed and will also incorporate some of the feedstock lipids into their cells and so this will be unavailable for degradation.

Sewage sludges, in particular, also contain hydrocarbon oils from drains. These will be virtually unattacked in digesters as bacterial degradation of hydrocarbons is essentially an aerobic process. Hydrocarbon oils and greases will merely form an oily scum on the digester contents.

Vegetable and other digester feeds also contain waxes and these will be little, if at all, degraded in the digester, but these are probably only a minor component of most digester feedstocks.

The lipids of digester feedstocks are, in the main, compounds of glycerol and long-chain fatty acids, but some lipids contain other, more complex, molecules as well as glycerol. The glycerol plus long-chain fatty acid lipids, 'glycerides', have been shown to be hydrolysed, and bacteria capable of this hydrolysis ('lipolysis') have been isolated from piggery-waste digesters (Hobson *et al.*, 1974) in numbers of 10^4–10^5/ml. Torien (1967) also isolated lipolytic bacteria from sewage digesters. The hydrolysis gives glycerol and a long-chain fatty acid, and the presence in digesters of bacteria which can ferment the glycerol has already been noted. The more complex glycerides will also undoubtedly be hydrolysed to give long-chain acids and it is the fate of these acids which will be discussed here.

The demonstration of the breakdown of long-chain fatty acids in digesters depends almost entirely on biochemical work as the bacteriology has not yet been elucidated.

The long-chain fatty acids can be 'saturated' or 'unsaturated'. The unsaturated acids do not have a full complement of hydrogen atoms attached to the carbon chain. In the digester these unsaturated acids are 'hydrogenated' by the bacteria to saturated acids (Heukelekian & Mueller, 1958). The long-chain acids are then degraded. The results of experiments (by, for example, Chynoweth & Mah, 1971 and Heukelekian & Mueller, 1958) show that this degradation is by a pathway in which two carbon atoms at a time are split from the acid chain to give acetic acid and a shorter long-chain acid. This is then repeated.

$$CH_3—CH_2 \} CH_2—CH_2—CH_2COOH + 2H_2O \rightarrow$$
long-chain acid

$$CH_3—CH_2 \} CH_2—COOH + CH_3COOH + \quad 4H$$
acetic acid \qquad hydrogen

This reaction results in the production of hydrogen. However, the reaction is not one which occurs readily. A bacterium has recently been isolated which will carry out this reaction, but it will do so only in the presence of a hydrogen-utilising bacterium, such as one producing methane from hydrogen and carbon dioxide. The removal of hydrogen 'pulls' the otherwise unfavourable reaction to the right of the equation (McInerney *et al.*, 1979). So far only the degradation of shorter-chain fatty acids (with up to seven carbon atoms in the chain) has been demonstrated. The long-chain acids from fats can contain up to 18 carbon atoms, but the mechanism should be the same whatever the chain-length of the acid. The acetic acid is also a substrate for the methanogenic bacteria. So the final result of the degradation of long-chain fatty acids is the formation of methane. With sewage-sludge digesters a major part of the methane generated can come from lipid in the feedstock because of the often high lipid content of the sewage sludge (Table 3.1).

MINOR BACTERIAL REACTIONS

As was pointed out before, some bacteria require vitamins and other bacteria can produce these. The production of vitamins, particularly the B vitamins, in digesting sludge has not been extensively investigated, but a production of vitamin B_{12} from digesting sewage was, and probably still is, a commercial proposition (Szemler & Szekely, 1969).

The production of sulphide from sulphate has already been mentioned, but other bacteria similar to those found in digesters can degrade sulphur amino acids to sulphide. Sulphide is used by some bacteria for production of their sulphur amino acids, but sulphide not used in this way will contribute to the small amounts of hydrogen sulphide found in digester gas. Other sulphide will combine with metals such as iron to give insoluble sulphides. The production of sulphide by the sulphate-reducing bacteria will obviously depend on the amount of sulphate in the feedstock. A mine water, containing high levels of sulphate, when added to a sewage digester increased numbers of sulphate-reducing bacteria from $6 \cdot 6 \times 10^3$ to $9 \cdot 5 \times 10^7/\text{ml}$ (Torien *et al.*, 1968). Since sulphide can inhibit methanogenesis (see later) the sulphate content of factory waste-waters or other 'unusual' digester feedstocks needs to be watched. Since sea water contains sulphate, excess sulphate might possibly pose problems in digesters fed with seaweeds.

Excessive production of hydrogen sulphide could lead to a rather obnoxious digested sludge and a gas which would have to be scrubbed before use. Sulphide incorporated into the bacteria does not smell. But other minor reactions can help in the reduction of odour found on digestion of excreta sludges. Acetic acid has rather a pungent smell and acetic acid in feedstocks is very much reduced in concentration during digestion. The volatile fatty acids higher than acetic, and especially the 'branched-chain' ones, have very unpleasant smells even in low concentration. Small amounts of these in sludges are either converted to methane and carbon dioxide or used to form components of bacterial cells. The small amounts of phenol and cresols, indole, skatole and lower fatty acids which contribute to the smell of faecal wastes are also destroyed during digestion (Van Velsen, 1977).

METHANE PRODUCTION

Although the overall reactions occurring in the final stage of digestion (the production of methane) are known, complete details of the bacteria and the biochemical mechanisms involved are not.

The main substrates for methanogenesis are acetic acid and hydrogen plus carbon dioxide, and acetic acid is usually regarded as the most important. Smith & Mah (1966) showed, for instance, that 73% of the methane in their domestic sewage digestion came from acetic acid. This kind of proportion might be expected from a consideration of the overall reactions occurring in the breakdown of carbohydrate and fats.

As discussed before, it is likely that the fermentation of carbohydrate is biased towards production of acetic acid, so the breakdown of carbohydrate can be represented as previously by:

$$C_6H_{12}O_6 + 2H_2O \rightarrow 2C_2H_4O_2 + 2CO_2 + 4H_2$$

The acetic acid and hydrogen products will then form methane as follows:

$$2C_2H_4O_2 \rightarrow 2CH_4 + 2CO_2$$

$$4H_2 + CO_2 \rightarrow CH_4 + 2H_2O$$

In the formation of methane from carbohydrate 66 % of the methane will have come from acetic acid and 33 % from hydrogen.

In each step of the long-chain fatty acid breakdown one acetic acid molecule and four hydrogens are formed as shown before. The acetic acid will give one CH_4, while the 4H will give $\frac{1}{2}CH_4$ as above, again in proportions of 2:1.

Hydrogen, either as a gas or on the previously-mentioned carrier molecules, is used for methane production. It is also used in the hydrogenation of the long-chain fatty acids, in the reduction of sulphate and nitrate and in the formation of bacterial cellular constituents. So not all hydrogen produced in fermentations is available for methane formation. This will increase the proportion of methane formed from acetic acid. Obviously, these equations are simplified and other acids than acetic will be formed to some extent, but the main argument remains. Animal excreta wastes always contain volatile fatty acids from fermentations occurring in the animal gut or in the voided faecal material. The acid concentrations, as will be shown later, can be very high and acetic is a major constituent of the acid mixture. So this again could contribute towards the preponderance of acetic acid as a precursor of methane. Hydrogen formed in these predigester fermentations would be given off as gas and lost.

However, although acetic acid is the major substrate for methane formation comparatively little is known about the numbers and types of bacteria responsible for this reaction. The hydrogen-utilising bacteria have proved more easy to isolate, even though culture of methanogenic bacteria in general is one of the more difficult bacteriological tasks. The absolute purity of cultures of methanogenic bacteria is often difficult to prove, so doubt may exist as to whether a reaction is being carried out by one bacterium alone or an association of two or more bacteria. Indeed, the production of methane from ethanol long thought to be carried out by a species of methane bacterium'was a few years ago shown to be a function of two types of bacteria making up what had been accepted as a pure culture of one bacterium (Bryant *et al.*, 1967).

Taken as a whole, the methanogenic bacteria are not easy to keep in laboratory cultures and many strains isolated by various authors have died, or have been described only in the initial cultures from a digester, having failed to grow when transferred to further culture media. There are undoubtedly many kinds of methanogenic bacteria, but how many is not known, neither is whether a similar variety of these bacteria exists in every digester nor whether the mixture varies with the feedstock or digester operating conditions.

The subject of the methanogenic bacteria is complex and there is not space here to discuss it in detail. The reader is referred to reviews and original papers such as the following: Wolfe, 1971; McBride & Wolfe, 1971; Bryant, 1976, 1979.

Just a few points will be made here.

Up until very recently (Zinder & Mah, 1979) all the methanogenic bacteria known would utilise hydrogen plus carbon dioxide, but only certain ones would use other substrates. The hydrogen-utilising *Methanobacterium formicicum* appears to be a common constituent of digester bacterial populations having been isolated from domestic sewage digesters at different times (e.g. Schnellen, 1947; Mylroie & Hungate, 1954; Buraczewski, 1964; Smith, 1966). Where counts have been done, numbers of 10^5–10^8 hydrogen-utilising bacteria per ml have been found in sewage digesters. *Methanobacterium formicicum* was identified as a principal hydrogen-utilising bacterium in piggery-waste digestions where numbers of 2×10^5 to 2×10^6 were found. The build up of this population paralleled the increase in methane production during the development of piggery-waste digestion from water-filled digesters (Hobson & Shaw, 1974).

This bacterium also uses formic acid as a substrate, as its name implies. Formic acid (H_2CO_2) is equivalent to H_2 plus CO_2 gases and can be formed in place of these gases in the production of acetic acid. If formic acid is formed in the digester fermentation it will be quickly used up, but it is possible that the digester pH as well as the effect of the methanogenic bacteria will tend to change reactions to the production of the gases rather than formic acid.

The *M. formicicum* is a Gram-negative, rod-shaped bacterium, but Gram-positive *Sarcina*-like bacteria (i.e. cocci in cubic groups of eight) which are capable of methane production from acetic acid have been isolated from digesters (Schnellen, 1947; Buraczewski, 1964; Smith, 1966; Mah *et al.*, 1978) but this is probably not the only acetate-utilising methanogenic bacterium. The authors, for instance, have not found these bacteria in acetic acid-utilising cultures from mesophilic piggery-waste digesters, a Gram-negative, slender rod being obtained. On the other hand, Gram-variable cocci mainly in twos have been predominantly isolated from poultry and cattle-waste digesters (Hobson, Bousfield & Summers, 1974, unpublished).

The formation of methane from propionic and butyric acids has been demonstrated in cultures obtained from diluted digester contents, and such cultures have enabled counts to be made of the numbers of bacteria in digesters able to carry out these reactions. However, such cultures contain a mixture of bacteria. By enrichment culture, stable mixed cultures containing two, or possibly three types of bacteria were isolated from piggery-waste digesters (Hobson *et al.*, 1974). Although, for instance, Buraczewski (1964) claimed to have isolated pure cultures of single species of bacteria which converted propionic and butyric acids to methane, the latest work has shown that degradation of these acids must almost certainly always be the result of the co-operative action of two bacteria acting in the manner already discussed in the section on lipid degradation. The degradation of propionic acid is the most unfavourably affected by hydrogen, so if methane formation from hydrogen is stopped, propionic acid will tend to accumulate in digesters. The fatty acid-degrading bacterium isolated by McInerney *et al.* (1979) and previously mentioned, converted fatty acids with even numbers of carbon atoms, which includes butyric acid, to acetic acid and hydrogen. Valeric and heptanoic acids which have five and seven carbons were degraded to acetic acid, hydrogen and propionic acid and the latter was not further metabolised.

However, whatever the types of bacteria involved it seems to be a general finding that the growth of the acetic acid-utilising bacteria is much slower than that of the hydrogen utilisers. The authors found that in continuous culture, mesophilic acetic acid-utilising bacteria were washed out at dilution rates about 0·01/h (a detention time of 100 h, see later) and incubation periods of eight to twelve weeks were needed in batch cultures for production of methane. This is similar to the nine weeks incubation used by Heukelekian & Heinemann (1939) for their incubations of acetic acid-containing dilutions of digester contents and to incubation times used by others. In tests on piggery-waste digesters, hydrogen- and butyric acid-utilising bacteria could be counted with cultures incubated for four weeks, but acetic-acid utilisers did not appear in this time.

McCarty (1966) found that conversion of acetic acid to methane became less efficient as the detention time of his digesters was brought below eight days, but that gasification of formic acid (equivalent to $H_2 + CO_2$) was still about 90 % efficient at a two-day detention time, the lowest tested. Ghosh & Klass (1978) found in the experiments previously described that a detention time of six and a half days was required for the safe operation of the second stage (conversion of acids to methane) of their two-stage digester and a minimum of three days was suggested. In their two-stage system Cohen *et al.* (1979) used a detention time of 100 h for the acid-degradation stage. The authors' experience that a piggery-waste digester was on the point of a drastic failure at a three-day detention time (see later) seems to be common to other digesters and related to the methanogenesis rates quoted.

The studies on methanogenesis in digesters thus show that acetic acid and hydrogen plus carbon dioxide are principal substrates, although butyric and propionic acids may be used if they are actually formed to any extent or present in the feedstock. And they also show that the conversion of acetic acid is more likely to be a rate-limiting reaction in digesters than the conversion of hydrogen.

The possibilities of methane production from the molecules which make up lignin has already been discussed, but such reactions cannot in general be very important in digestion.

FACTORS AFFECTING THE RATES OF GROWTH AND ACTIVITIES OF DIGESTER BACTERIA AND INTERACTIONS AMONGST THE BACTERIA

In a continuous culture of one bacterium, with one growth-rate-limiting substrate and all substrates in solution and composed of small molecules such as sugars or amino acids, then (it will be shown later) the factor determining the growth rate of the bacteria will be the rate of flow of the medium up to a certain value which corresponds with the maximum possible growth rate of the bacterium. So within this limit the culture can be run at any rate. In a mixed culture such as a digester the maximum growth rates of many bacteria have to be taken into account as have the rates of degradation of many complex substrates. These bacteria are not growing in isolation in the mixture, they are interdependent and so the rate of reaction of one or a group of bacteria can control the rates of reaction of many others, and the elimination of one bacterium or group may stop all the reactions in the digestion. Some of these interactions controlling digestion have already been described, other interactions will be discussed here.

It was previously mentioned that digesters usually contain aerobic and facultatively anaerobic bacteria derived from the feedstock. The structure of digesters is designed to eliminate ingress of air to the digesting sludge, but some oxygen, dissolved in feedstock water perhaps, or from leaks in the system, may get into the digester.

Faecal wastes contain large numbers of aerobic or facultatively-anaerobic bacteria. Studies on domestic sewage or animal-waste digesters show that a large proportion of the digester floras are bacteria capable of aerobic metabolism; in fact early investigators of domestic sewage digesters isolated little but such bacteria in numbers of the order of 10^6/ml (Pohland, 1962). Later workers using specialised techniques have demonstrated the importance and numbers of the strictly anaerobic bacteria. However, these workers have also shown that the aerobic and facultative bacteria isolated by the earlier workers do exist in digesters in the kind of numbers they found. The suggestion from detailed counts is that the facultative and aerobic bacteria in domestic digesters are in number less by a factor of 10 or 100 than the anaerobic bacteria (Torien *et al.*, 1967; Mah & Sussman, 1968). On the other hand the facultative bacteria in piggery-waste digesters were found to be in very much the same numbers as the anaerobes (Shaw, 1971; Hobson & Shaw, 1971, 1974). The higher proportion of facultative bacteria in the piggery-waste digesters compared with the domestic digesters could be explained by the very large numbers of such bacteria in the feedstock. These numbers, at $2\cdot4 \times 10^8$/ml of a piggery-waste slurry of about 4% TS, were higher than the numbers in the digester itself, 5–7 $\times 10^6$/ml. The figures indicate that a large number of the bacteria died off in the digester, nevertheless very many remained viable, if not actually actively metabolising, in the digester sludge, as discussed before.

These bacteria, while not engaged in the main degradative reactions of digestion could play some role in general sugar fermentations, but they are not necessary for these. Where they most likely play an important part in the ecology of the digester is as scavengers of oxygen, reducing the system to an Eh suitable for growth of the strict anaerobes and in particular the methanogenic bacteria. In this the bacteria play a similar role to aerobically-metabolising bacteria in the rumen or in muds or stagnant waters. In farm-waste digesters the aerobic and facultative bacterial population can metabolise large amounts of oxygen. A leak of air into a digester need not, then, lead to an inflammable mixture of methane and oxygen in the digester gas. In an experimental 13-m^3 digester an initially undetected hole in a sludge-heater circulation pipe allowed air to be drawn into the digester. No apparent decrease in gas production had been noted, but the gas became non-flammable in the heater boiler. Analysis of the gas in the digester head-space soon after this showed that it was 5·2% methane, 23·1% carbon dioxide and 71·7% nitrogen, with no oxygen. The oxygen in the air leaking in had been metabolised by the aerobic and facultative bacteria. However, it seemed that the bacteria had almost reached their limit of oxygen-metabolising capacity because five hours later the gas in the head-space had almost the same ratio of oxygen to nitrogen as in air and both methane and carbon dioxide were very low. At this point the hole in the pipe was found and repaired. The oxygen in the head-space gas was metabolised in a few hours after the leak had been stopped but gas production was very small for some days and about eight days at normal loading rates was needed to return gas production to normal and to flush all the residual nitrogen out of the system. In the first day or two the methane to carbon dioxide ratio was 35:65, next day 40:60, two days later 54:46 and three days after this 68:32 and the digester was functioning normally (Hobson *et al.*, 1974). Thus, although there is a large capacity for oxygen uptake, too much oxygen can raise the Eh of the digester contents to a level at which the fermentative and methanogenic bacteria cannot metabolise, but even after this, reduced conditions can again be established.

A digester usually runs naturally at a pH of 7 or just over, about 7·2. This pH is about optimum for the methanogenic bacteria. Ghosh & Klass (1978) used a pH of 7·2 for optimum running of the methanogenic stage of a two-stage laboratory-scale digester and Cohen *et al.* (1979) ran their methanogenic stage at a pH of 7·8, which was the uncontrolled steady-state pH of the reactor. Hamer & Borchardt (1969) found the optimum pH for methane production from acids was 7·05–7·20. These pH values agree with a pH range for growth of 6·5–7·7 found by Smith & Hungate (1958) for a hydrogen-utilising methanogenic bacterium found in digesters and the rumen. However, such a high pH may not be optimum for the fermentative bacteria, but the optimum pH for these may depend on the substrates being utilised. Ghosh & Klass (1978) found that the optimum pH for the fermentative acidification stage of their small-scale two-stage digestion of sewage sludge or glucose was 5·7–5·9, and Cohen *et al.* (1979) ran the acidification stage of a two-stage digester using glucose at a pH of 6·0. These results suggest that the acid-forming bacteria in a normal, single-stage digester are working well away from their optimum pH. However, this may not always be so. It has already been pointed out that the sludge used by Ghosh & Klass appeared to contain very little cellulose and the other substrate quoted above was the sugar, glucose. The optimum pH for breakdown of cellulose in grass and other fibres in the rumen is about 6·5 or rather higher; cellulose fermentation is decreased at a pH of 6. Since the digester bacteria are similar to rumen ones, their pH optimum might be expected to be similar. The optimum pH for cellulolytic enzymes from a variety of sources is about 6·7. Thus a digester pH of 7·2 may not be as far away from the optimum for the cellulose-fermenting bacteria as is sometimes suggested. The starch-digesting anaerobic bacteria, however, seem to have a lower pH optimum than the cellulolytic bacteria, one rumen bacterium having optimum production of starch-hydrolysing enzyme (α-amylase) at pH 6·1 (Hobson & Summers, 1967) and the optimum pH of the bacterial amylases is generally low (Hobson & MacPherson, 1952).

There are two ranges of temperature over which the complex populations of bacteria concerned in anaerobic digestion will grow. These are the 'mesophilic' and 'thermophilic' ranges. Different bacteria grow at these two ranges which are from about 5 °C to 45 °C or 50 °C and from about 55 °C to about 70 °C. In either case the bacteria grow faster as the temperature is increased from the lower limit to a temperature of maximum growth rate, and then rapidly decrease in growth rate as this temperature is passed. For example, in the mesophilic range the bacteria grow fastest at about 44 °C, but a few degrees higher their activity has decreased considerably. However, although consideration purely of bacterial growth suggests that whichever of the ranges is chosen for digester running a temperature near the top of the range should be chosen, practical considerations can dictate a lower working temperature as will be seen later. Similarly, although the thermophilic bacteria generally grow faster than the mesophilic bacteria, practical considerations are against the use of thermophilic digesters in many cases.

However, whatever the actual temperature of operation once the bacteria have been adapted to a particular temperature this should be controlled as strictly as possible as a lowered efficiency can result if bacteria are subjected to sudden temperature changes, even of 5 °C or so.

It has already been explained how the digester bacteria form a consortium carrying out a series of reactions and how the bacteria are mutually interdependent for nutrients and subsidiary growth factors. In the event of one link in the chain of bacteria and reactions failing the whole system can collapse. It is probably unlikely in practical digester operation that a chemical specifically poisoning one group of bacteria or inhibiting one reaction will get into the feedstock, although chlorinated hydrocarbons such as chloroform or carbon tetrachloride are specifically inhibitory in very small amount for the methanogenic bacteria. Some antibiotics are specific for one type of bacterium, but others inhibit a wide range of bacteria. Antibiotics in sufficient concentration to inhibit the digester bacteria are unlikely to be found in feedstocks, except, perhaps, those formed from farm-animal excreta. Antibiotics used to be commonly added in relatively large amount to farm-animal feedstuffs to promote animal growth. However, because of the dangers of producing pathogenic bacteria which are resistant to the antibiotics, their use in animal feedstuffs has been discouraged or banned in many countries, particularly Britain. One antibiotic which is allowed and which is now becoming a more common addition to animal feeds is monensin (sold under trade-names). This is used specifically to decrease methane formation in the rumen of cattle and waste from cattle with monensin added to the feed might affect digestion. The uncontrolled use of antibiotics in attempts to stop an outbreak of some disease in a group of farm animals could be a possible hazard in operation of farm-waste digesters. Turnacliff & Custer (1978) reported that an experimental digester in America was unsuccessful in start-up and running because 40 g of Lincomycin was routinely added to each tonne of pig feed used on the farm, to prevent swine dysentery. Such feeding would be impossible in Britain, but addition of Lincomycin to pig feeds can be done for short periods under veterinary control to cure an outbreak of dysentery. In such cases it would appear best to divert the waste from the digester while antibiotics are being used. The maximum concentration of antibiotics that could be tolerated by digester bacteria is not, at present, known. But bacteria in mixed populations can adapt to become resistant to, or degrade, many antibiotics, so antibiotics in small amount continuously added to a digester with the feedstock will probably not affect its running. Various additives to poultry feeds, such as arsenicals and zinc bacitracin did not affect running of the authors' digesters.

Undesirable substances which may get into digester feedstocks are more likely to be general bacterial poisons. Wastes from chemical plants getting into town sewers have been known to poison anaerobic digesters, for instance pentachlorophenol in a concentration of 1·4 mg/litre was shown to be responsible for failure of two sewage digesters near London some years ago (Drew & Swanwick, 1962). With the increasing investigation of digestion as a means of gas production from various vegetable feedstocks more possibilities of digester inhibition may occur. For instance the oils of citrus fruit peels can be inhibitory and McNary *et al.* (1951) found that D-limonene from the peels inhibited digestion of citrus fruit wastes. Some of the oils and resins in woods are also bacterial inhibitors, particularly those from some pines.

Detergents get into sewage and so into sewage-sludge digesters, and they may get into some agricultural-waste digesters, e.g. from the milking parlour washings on dairy farms. Little is known about the effects of detergents on farm-waste digesters, but no problems have so far been reported, probably because any detergents used in cleaning are considerably diluted with water used to wash floors and pipe lines and are again diluted on admixture with the animal wastes. Some experiments have been done on the effects of detergents on sewage-sludge digestion, for example by Bruce *et al.* (1966). They said that occasional reports had been made of reduced activity in digesters which might have been due to high concentrations of detergents in the sewage. Some instances of high detergent concentrations could be due to industrial discharges, but Bruce *et al.* quoted results of experiments on sewage from a non-industrial area which had concentrations of detergent sufficient to partially inhibit digestion. Extensive laboratory digester experiments showed that primary alcohol sulphate detergents were degraded anaerobically and even up to 4·2 % of sludge solids showed no more than a slight retardation of gas production when they were first added to the sludge, but eventually they slightly increased gas yield. Sodium alkane sulphonates were not degraded anaerobically and inhibition of digestion was caused at concentrations above 2 % of the sludge solids. Both 'hard' and 'soft' alkyl benzene sulphonate detergents, which can be distinguished by aerobic degradability, were resistant to anaerobic degradation. Concentrations up to about 1·5 % of sludge solids had little effect, between 1·5 and 2 % digestion might be erratic, and over 2 % more serious inhibition generally occurred. The ratio of detergent to sludge solids appeared to be a more important factor than the total concentration in the wet sludge. Since dairy-cow-waste digesters will usually be running on a high-solids input (*c*. 8–10 %, see later) this fact is of importance in assessing possible effects of the dairy washings.

'Heavy metals' (copper, cadmium, etc.) are inhibitory to most microbial processes, and these can get into domestic sewage from factory wastes, and into farm wastes; the latter mainly because of the common practice of adding copper to pig feeds as an animal growth stimulant. However, these metals, with the exception of chromium, form very insoluble sulphides and so in the digester where hydrogen sulphide is being produced they are likely to be precipitated as the sulphide. Only the metal actually in solution as an 'ion' can affect the bacteria. In piggery-waste digesters copper has not been found to be a problem. It is almost entirely removed from action on the bacteria as it is in the form of copper sulphide in the solids of the digester contents. The copper sulphide is formed from the soluble copper sulphate in the pig feed by reduction or by hydrogen sulphide produced by bacterial action in either the pig gut or the digester. The precipitation of metals to differing extents is most likely responsible for different concentrations of metals being given as toxic to digestion. The amount of hydrogen sulphide produced will vary, for example with the sulphate content, as with the amount of protein (and thus sulphur-containing amino acids) in the feedstock. Mosey *et al.* (1971) discussed factors affecting the availability of heavy metals to inhibit digestion. Amongst these was sulphide, but the carbon dioxide in solution in the digester liquid could also precipitate some metals—zinc, cadmium, copper and lead—as carbonates. Because of effects on the solubility of carbonates the pH value of the digester contents affects toxicity of the metals, and amongst procedures suggested for restoring a digester failing because of metal toxicity was the addition of sodium carbonate to raise the digester pH to about 8. Lime was not recommended because of the precipitation of the lime as calcium carbonate, which itself can cause problems, and raising the pH with ammonia carries with it the possibility of causing ammonia toxicity. The preferred method was addition of sodium sulphide to precipitate the metals as sulphides. To prevent the possibility of hydrogen sulphide being given off in amounts which could be dangerous to operatives it was recommended that the pH of the digester contents should, if necessary, be raised to 7–7·5 and the sodium sulphide be dissolved in sludge of pH 7·5 or in an alkaline solution. (The pH of an inhibited digester may fall below the usual value of 7 or over because of accumulation of acids after inhibition of methanogenesis.)

However, sulphide itself can be toxic to digestion and concentrations of soluble sulphide (as dissolved hydrogen sulphide or sodium or potassium sulphide for instance) in excess of 100 mg/litre are generally regarded as toxic. In the previously quoted paper the authors suggested that although sodium sulphide addition was to be preferred on account of the speed with which it will react with the heavy metals, in the absence of adequate measurements of sulphide concentration iron (ferrous) sulphate should be the reagent added. The sulphate is reduced to sulphide which precipitates the heavy metals, but any excess sulphide formed will form insoluble iron sulphide and so will not be toxic.

McCarty (1964) earlier suggested sulphide precipitation as a means of preventing heavy metal toxicity in digesters. He also pointed out that while sodium, potassium, calcium and magnesium ions in solution of stimulatory to digestion in concentrations of about 75–400 mg/litre (depending on the metal), in concentrations in the 1000–5000 mg/litre range they are somewhat inhibitory, and in concentrations for magnesium of 3000 mg/litre and for the other metals of 8000–12 000 mg/litre they are strongly inhibitory. He also pointed out that sulphide in solution as hydrogen sulphide is driven off with the digester gas as gaseous hydrogen sulphide. The amount given off depends to some extent on the pH of the digester (approximately 7) but at any pH the amount given off depends on the rate of gas production. So a vigorously gassing digester can apparently tolerate more sulphide than a poorly gassing one. He also suggested that while about 100 mg of sulphide/litre is generally toxic to digestion, digesters can be acclimatised to tolerate up to 200 mg/litre if the sulphide concentration gradually increases. Concentrations above 200 mg/litre are always toxic.

The long-chain fatty acids formed from lipids can be inhibitory or stimulatory to methanogenic and fermentative bacteria depending on acid and concentration, as was shown by experiments on rumen bacteria by Henderson (1973) and Prins *et al.* (1972). The methanogenic bacterium, *Methanobacterium ruminantium*, used in these experiments has also been found in digesters, and oleic acid was particularly inhibitory at more than 50 mg/litre. McCarty (1964) also found that the long-chain fatty acid, oleic, inhibited anaerobic digesters in concentrations of 500 mg/litre. However, as in the case of heavy metals and sulphides, long-chain fatty acids can be precipitated from solution as calcium salts and in the presence of sufficient calcium 2000–3000 mg/litre of oleic acid could easily be tolerated in digesters. Most long-chain fatty acids in domestic waste are in the form of calcium salts ('soaps').

Ammonia itself, although necessary in digesters as a nitrogen source for the bacteria, can be toxic to digester bacteria when present in excess. Hobson & Shaw (1976) showed that *Methanobacterium formicicum* from piggery-waste digesters was inhibited by ammonia concentrations of 3000 mg/litre at pH 7·1 and its growth was completely stopped by 4000 mg/litre. This accords with McCarty's (1964) findings that at any pH, 3000 mg ammonia nitrogen/litre was inhibitory to anaerobic digesters and at pH 7·4 and above concentrations of 1500–3000 mg/litre of ammonia nitrogen were inhibitory. The effect of pH is caused by the fact that the relative concentrations of ammonia and ammonium ions in solution are affected by pH. Ammonia in gaseous form or as dissolved ammonia gas is NH_3, but when combined in the form of salts it is as the ammonium ion NH_4^+. In solution the ammonium dissociates, $NH_4 \rightleftarrows NH_3 + H^+$, in a 'reversible' reaction. When the hydrogen ion (H^+) concentration in the solution is sufficiently high then the reaction is pushed to the left and ammonium ions are the main constituents of the mixture. At higher pH values ammonia gas in solution is the main constituent. Ammonia is more inhibitory than ammonium ions. In practice, however, total 'ammonia' or 'ammonia nitrogen' is measured and this includes true ammonia and ammonium, so the concentrations previously given were for ammonia as measured. Some farm-animal wastes can contain very high concentrations of ammonia formed in part by microbial action on various nitrogenous compounds while the animal excreta are standing in or moving slowly through collecting systems and an example of inhibition of poultry-waste digestion of which the most likely cause is excess ammonia will be given later. In these experiments the digester was running for many months at very high ammonia concentration and inhibition was gradual. Van Velsen (1979) has suggested on the basis of batch experiments that the digester bacteria can acclimate to ammonia concentrations up to about 5000 mg/litre and that the successful digestion of piggery waste, where ammonia concentrations are 2–3000 mg/litre, is due to the acclimation of the bacteria. The reported inhibitions of sewage digesters at ammonia levels above about 1500 mg/litre could be due to the bacteria usually being presented with a feedstock of lower ammonia concentration. The effects of ammonia are complex. On the other hand, experiments on rumen metabolism (Mehrez *et al.*, 1977) have shown that a certain 'steady-state' concentration of ammonia seems to be required for maximum bacterial activity and lack of ammonia is later suggested as a reason for low gas-production from diluted cattle wastes.

The volatile fatty acids, acetic, propionic and butyric, may be inhibitory to methanogenesis as well as being substrates for this reaction. However, the inhibition by acetate or butyrate is not significant. Hobson & Shaw (1976) found that growth of *M. formicicum* was not inhibited by concentrations of up to 10 000 mg/litre of acetic or butyric acids (measured as acetic acid), and McCarty & McKinney (1961) had earlier observed that acetic acid in high concentration did not inhibit digestion as a whole. On the other hand, the experiments of Andrews (1969) suggested that propionic acid was inhibitory to laboratory digestions and Hobson & Shaw (1976) showed that the lowest concentration of propionic acid tested (1000 mg/litre, as acetic) partly inhibited growth of *M. formicicum*. The bacteriological experiments were carried out at a pH of 7·1, while the other tests were also at normal digester pH. While pH may have some effect on observed toxic concentrations of acids because of changes in the relative concentrations of ionised and unionised forms of the acid, pH itself, as a reflection of conditions more acid or alkaline than the normal growth ranges of the bacteria, can be said to be inhibitory. Acid conditions can be brought about in a digester by a sudden, large addition of an easily fermented carbohydrate when the production of acids can be greater than the removal of acids by methanogenesis. Acids then build up in concentration and can overcome the buffering action of the digester contents (see below).

However, it should be noted that while a large and sudden addition of some substance, such as carbohydrate, over and above that normally present in the feedstock, can unbalance the metabolic pathways and cause digester failure, if small but increasing amounts of that same substance are added daily to a continuous-flow digester then over some days or weeks the balance of the bacterial population will change and the digestion can be 'adapted' to metabolise the large daily input of substance which added suddenly would cause digester failure. Digesters adapted to farm wastes can tolerate very large concentrations of acids in the feedstocks because the acids are rapidly metabolised and the 'steady-state' concentration of acid in the digester is low (see later). The acid is mainly acetic acid, which is easily metabolised, but even the less-readily degradable propionic acid, in inhibitory concentration in the feedstock, is reduced in concentration sufficiently to be non-inhibitory.

The reduction in concentration of acids also reduces any tendency for the digester to fall to an acid, and thus inhibitory, pH. The pH of the digestion is controlled by 'buffering' action. Hydrochloric acid (HCl) forms, obviously, an acid solution (i.e. the pH is low). Sodium hydroxide solution is strongly alkaline (high pH). Mixed together they form sodium chloride, which is neutral in pH. Similarly, in digesters, ammonium hydroxide (ammonia in water), which is alkaline, neutralises the acidic acetic acid. In the digester are acids and alkalis, such as the acetic acid and ammonia, and carbon dioxide in solution (carbonic acid), various metal salts and amino acids (which have both acidic and alkaline properties) and these various factors tend to neutralise each other and thus keep the pH of the digester contents at approximately 7. There is usually an excess of ammonia and this can neutralise excess acid (formed in the digester or added as just described), so this acid will not alter the digester pH significantly until such a high concentration is reached that the neutralising or 'buffering' action of the ammonia and other ions is overcome. In practice, then, digesters and especially farm-waste ones, can contain very high concentrations of acetic acid without the pH being adversely affected. The buffering action of digester contents can be measured as the 'alkalinity'.

The various inhibitions or stimulations of digester bacteria are not generally straightforward, and stimulation or inhibition may be increased or decreased by interactions of the various chemicals. Bacteria are also adaptable and either the bacteria present can change to overcome some inhibitory compound, or new species of bacteria resistant to the inhibitor can grow up to replace the existing ones. But this can only happen if changes in feedstock or operating conditions are gradual. Some conditions cannot be adapted to and these are the ones to be avoided in digester running.

CONCLUSIONS

In the preceding sections and chapters the microbiology and biochemistry of anaerobic digestion have been briefly reviewed. The biochemical pathways are complex and the mixture of bacteria involved is also complex, and neither the reactions nor the bacteria have been fully characterised. Nevertheless, the bacteria and the reactions occur in natural habitats and given suitable conditions the population of bacteria will develop in man-made habitats and form a largely self-regulating system. Even if the details of the reactions are not known, enough has been found out about the behaviour of the system as a whole to enable digesters to be run, or planned to run, with various feedstocks and to enable the activities of the systems to be more or less optimised.

In the next chapters attention will be turned from the microbiology of the process to practical applications, actual or proposed. In many cases large-scale plant has not yet been built and run, but small-scale experiments have determined the parameters for digestion of the feedstock. Large-scale digestion of wastes other than domestic sewage is a subject on which many people are now working and new digesters are continually being built and tested. This activity has increased over the last two or three years and is still increasing, so by the time this text appears some projects may have reached a more advanced state than is described here, and other projects not known of at the moment may have achieved success. Such advances to successful plants will be achieved by detailed study and overcoming of the engineering problems associated with feedstock handling and digester running rather than by any startling 'breakthrough' in the basic microbiology and digestion parameters described here. The basis for successful digesters is known; successful application in many areas should soon be achieved.

Digestion, as has already been explained, can be used for two main purposes; for energy production or pollution control. The emphasis at the moment is largely on energy production, but in general both aspects go together and most effective gas production from a waste usually gives most effective pollution control. Monetary or energetic economics applied for the purpose of generating the maximum amount of cheap energy also minimise costs of pollution control. But while pollution control remains a basic reason for digesting some feedstocks, other feedstocks are in effect non-polluting as without the need for producing them as energy sources they would not be there, or would be used or disposed of for other purposes; such are the energy crops. However, in all cases there remains a digested sludge residue. This has to be got rid of in some way and if it can form a useful fertiliser, a source of animal or human feed or a further source of fuel, for instance, then this not only solves one of the problems of running a digester plant but further helps in the economics of the process and the overall benefits of the process to man and the environment. So in any survey of digestion the ultimate fate of the digester effluent must be considered— this will be done in the following chapters.

REFERENCES

ANDREWS, J. F. (1969). *J. Sanit. Eng. Div., Proc. Am. Soc. Civ. Eng.* **95**, 95.

BRUCE, A. M., SWANWICK, J. D. & OWNSWORTH, R. A. (1966). *J. Inst. Sew. Purif.* **5**, 3.

BRYANT, M. P. (1976). In: *Seminar on Microbial Energy Conversion*, Germany (ed. H. G. Schleigel) E. Goltz KG, Göttingen, Germany. p. 107.

BRYANT, M. P. (1979). *J. Anim. Sci.* **48**, 193.

BRYANT, M. P., TZENG, S. F., ROBINSON, I. M. & JOYNER, A. E. JR. (1971). In: *Anaerobic Biological Treatment Processes. Adv. Chem. Ser.* **105**, 23.

BRYANT, M. P., WOLIN, E. A., WOLIN, M. J. & WOLFE, R. S. (1967). *Arch. Mikrobiol.* **59**, 20.

BURACZEWSKI, G. (1964). *Acta Microbiol. Pol.* **13**, 321.

BUSWELL, A. M. & NEAVE, S. L. (1930). *Ill. State Water Surv. Bull.* **30**.

CHUNG, K. T. (1972). Abstr. AGM Am. Soc. Microbiol. p. 64.

CHYNOWETH, D. P. & MAH, R. A. (1971). In: *Anaerobic Biological Treatment Processes. Adv. Chem. Ser.* **105**, 41.

COHEN, A., ZOETEMEYER, R. J., VAN DEURSEN, A. & VAN ANDEL, J. G. (1979). *Water Res.* **13**, 571.

DREW, E. A. & SWANWICK, J. D. (1962). *Publ. Wks. Mun. Serv. Cong.*, p. 1.

GHOSH, S. & KLASS, D. L. (1978). *Proc. Biochem.* **13**, 15.

HAMER, M. J. & BORCHARDT, J. A. (1969). *J. Sanit. Eng. Div., Proc. Am. Soc. Civ. Eng.* **95**, 907.

HARKNESS, N. (1966). *J. Proc. Inst. Sew. Purif.* p. 542.

HEALY, J. B. & YOUNG, L. Y. (1979). *Appl. Environ. Microbiol.* **38**, 84.

HENDERSON, C. (1973). *J. Appl. Bacteriol.* **36**, 187.

HENDERSON, C. (1980). *J. Gen. Microbiol.* **119**, 485.

HEUKELEKIAN, H. & HEINEMANN, B. (1939). *Sew. Wks. J.* **11**, 426.

HEUKELEKIAN, H. & MUELLER, P. (1958). *Sew. Ind. Wastes* **30**, 1108.

HOBSON, P. N. (1976). In: *Proc. Int. Solar Energy Conf.* UK-ISES, London.

HOBSON, P. N. (1979). In: *Microbial Polysaccharides and Polysaccharases* (eds. R. C. W. Berkeley, G. W. Gooday and D. C. Ellwood) Academic Press, London and New York. p. 377.

HOBSON, P. N., BOUSFIELD, S. & SUMMERS, R. (1974). *Critical Reviews in Environmental Control* **4**, 131.

HOBSON, P. N. & MCPHERSON, M. J. (1952). *Biochem. J.* **52**, 671.

HOBSON, P. N. & SHAW, B. G. (1971). In: *Microbial Aspects of Pollution* (eds. G. Sykes and F. A. Skinner) Academic Press, London and New York, p. 103.

HOBSON, P. N. & SHAW, B. G. (1973). *Water Res.* **7**, 437.

HOBSON, P. N. & SHAW, B. G. (1974). *Water Res.* **8**, 507.

HOBSON, P. N. & SHAW, B. G. (1976). *Water Res.* **10**, 849.

HOBSON, P. N. & SUMMERS, R. (1967). *J. Gen. Microbiol.* **47**, 53.

HOBSON, P. N. & SUMMERS, R. (1979). In: *Techniques for the Study of Mixed Populations* (eds. D. W. Lovelock and R. Davies) Academic Press, London and New York, p. 125.

HUNGATE, R. E. (1950). *Bact. Rev.* **14**, 1.

IANNOTTI, E. L., KAFKEWITZ, D., WOLIN, M. J. & BRYANT, M. P. (1973). *J. Bacteriol.* **114**, 1231.

KIRSCH, E. J. (1969). *Dev. Ind. Microbiol.* **10**, 170.

KOTZÉ, J. P., THIEL, P. G., TORIEN, D. F., HATTINGH, W. H. J. & SIEBERT, M. L. (1968). *Water Res.* **2**, 198.

LATHAM, M. J. & WOLIN, M. J. (1977). *Appl. Environ. Microbol.* **34**, 297.

MCBRIDE, B. C. & WOLFE, R. S. (1971). In: *Anaerobic Biological Treatment Processes. Adv. Chem. Ser.* **105**, 11.

MCCARTY, P. L. (1964). *Publ. Wks.* **95**, 91.

MCCARTY, P. L. (1966). *Dev. Ind. Biol.* **7**, 144.

MCCARTY, P. L. & MCKINNEY, R. E. (1961). *J. Water Pollution Control Federation* **33**, 223.

MCINERNEY, M. J., BRYANT, M. P. & PFENNING, N. (1979). *Arch. Mikrobiol.* **122**, 129.

MCNARY, R. R., WOLFORD, R. W. & PULTON, V. P. (1951). *Food Technol.* **5**, 319.

MAH, R. A., SMITH, M. R. & BARESI, L. (1978). *Appl. Environ. Microbiol.* **35**, 1174.

MAH, R. A. & SUSMAN, C. (1968). *Appl. Microbiol.* **16**, 358.

MAKI, L. W. (1954). *Antonie Van Leeuwenhoek* **20**, 185.

MEHREZ, A. Z., ØRSKOV, E. R. & MCDONALD, I. (1977). *Brit. J. Nutr.* **38**, 437.

MOSEY, F. E., SWANWICK, J. D. & HUGHES, D. A. (1971). *J. Water Pollution Control Federation* **6**, 2.

MYLROIE, R. L. & HUNGATE, R. E. (1954). *Can. J. Microbiol.* **1**, 55.

PFEFFER, J. T. (1978). *Proc. Biochem.* **13**, 6.

POHLAND, F. C. (1962). *Eng. Ext. Bull.*, Purdue Univ. USA, **20**, 583.

PRINS, R. A., VAN NEVEL, C. J. & DEMEYER, D. I. (1972). *Antonie Van Leeuwenhoek* **38**, 281.

ROBBINS, J. E., ARNOLD, M. T. & LACHER, S. L. (1979). *Appl. Environ. Microbiol.* **38**, 175.

SCHNELLEN, C. G. T. P. (1947). Dissertation, Technische Hoogeschool, Delft.

SHAW, B. G. (1971). Ph.D. Thesis, University of Aberdeen.

SIEBERT, M. L. & TORIEN, D. F. (1969). *Water Res.* **3**, 241.

SMITH, P. H. (1966). *Der. Ind. Microbiol.* **7**, 156.

SMITH, P. H. & HUNGATE, R. E. (1958). *J. Bacteriol.* **75**, 713.

SMITH, P. H. & MAH, R. A. (1966). *Appl. Microbiol.* **14**, 368.

SUMMERS, R. (1978). In: *Report on Straw Utilisation Conf.*, MAFF, p. 84.

SUMMERS, R. & BOUSFIELD, S. (1980). *Ag. Wastes* **2**, 61.

SZEMLER, L. L. & SZEKELY, A. D. (1969). *Proc. Biochem.* **4**, 12.

THIEL, P. G. (1969). *Water Res.* **3**, 215.

TORIEN, D. F. (1967). *Water Res.* **1**, 147.

TORIEN, D. F., SIEBERT, M. L. & HATTINGH, W. H. J. (1967). *Water Res.* **1**, 497.

TORIEN, D. F., THIEL, P. G. & HATTINGH, W. H. J. (1968). *Water Res.* **2**, 505.

TURNACLIFF, W. & CUSTER, M. (1978). Final Report to the State of Colorado Office of Energy Conservation, Contract *No.* OEC—00003.

VAN VELSEN, A. F. M. (1977). *Neth. J. Agric. Sci.* **25**, 151.

VAN VELSEN, A. F. M. (1979). *Water Res.* **13**, 995.

WAITE, R. & GORROD, A. R. N. (1959). *J. Sci. Fd Agric.* **10**, 317.

WOLFE, R. S. (1971). *Adv. Microbiol. Physiol.* **6**, 107.

YOUNG, J. C. & McCARTY, P. L. (1969). *J. Water Pollution Control Federation* **41**, 160.

ZINDER, S. H. & MAH, R. A. (1979). *Appl. Env. Microbiol.* **38**, 996.

CHAPTER 4

Types of Digesters: Theoretical Aspects and Modelling of Digester Systems and Deviations from Theory

There are a number of ways in which digestion can be carried out even with one type of feedstock, and considering various types of feedstock the number of ways can be multiplied. The question of which method to use may be answered in principle by consideration of what is known of the methods of laboratory or industrial fermentation, but the successful application of these methods is generally a matter of trial and testing of small- and then larger-scale plants. Complete design of a plant depends on knowledge of all factors involved, and in many digester situations these are not known. Nevertheless, theory and modelling can have some place in digester design. The theories of bacterial growth and running of batch and continuous cultures may give some help in considering which of two systems to adopt or what basic design features are needed. These theories lead to what one might call 'biological models'.

In deciding whether a digestion plant will produce useful amounts of usable energy, or whether the pollution control required will be energy-consuming or -producing, what might be termed 'engineering models' can be helpful. This type of model takes the basic results of the small-scale investigation on gas produced from a feedstock and the running parameters required to produce this gas and calculates from basic principles of energy losses in the plant, conversion efficiencies of gas to other forms of energy, the energy required to run the plant, and so on, and whether at all, or on what scale, digestion of such a feedstock would be energetically useful. This type of model is then taken a stage further in calculating the actual design parameters for the individual items of the plant when built, but this is a matter more of practical engineering than modelling as such.

The basic engineering model just described may be taken a step further when production of a feedstock specially for digestion, such as an energy crop, is proposed. Here the energy cost of producing the feedstock may be calculated to see if overall the process is energetically in positive balance.

The engineering model may be linked with a 'monetary model' in which

52

the overall economics of the plant or complete system are considered. This type of model is distinct from the actual costing of a digestion plant in a normal contracting way. It is used in attempting to assess the probable costs of energy production by digestion considered on a single plant or a country-wide basis, usually to see whether these costs bear any relationship to costs, actual or predicted, of other forms of energy.

Some of these models can be made fairly exact; some, because of lack of defined data to put into them, are very vague. In the case of the biological models the substrates and feedstocks make application of theories difficult, and there are practical constraints placed on these and other models, so various aspects of theory and modelling will be considered in the following sections.

BIOLOGICAL MODELS

Systems for Use with 'Wet' Feedstocks

As previously explained, microbes require water and all feedstocks for digestion must contain water. Many of the feedstocks are actually free-flowing, or fairly free-flowing, waters with material in solution and/or suspension. These must be treated in some form of liquid culture system. Others may be sludges of very high solids content which flow with difficulty. Biological (presence of inhibitory substances, etc.) or engineering (transport of the material, etc.) reasons may dictate that these are best diluted with water and, again, treated in a liquid culture. Some feedstocks are in effect solids, although containing water (vegetable matter, etc.); it may be that these can be digested as a solid mass. Such 'dry' digestions will be considered later, but in the following sections various possible systems for digesting wet feedstocks will be considered. One usual aspect of microbiological practice will not be considered, and that is sterility of feedstock and processing. Digestion is a mixed-bacteria culture and as such is largely self-regulating and organisms not 'required' for the process are kept to low numbers or completely inhibited by the 'active' bacterial population or the process conditions. A micro-organism-free feedstock is not required for a successful digestion. The engineering problems and monetary costs of running digesters under the sterile conditions required for a factory production of a single-cell protein or metabolic product from a pure culture of one organism in a defined culture medium would be immense and out of all proportion to the value of the digestion as energy-producer or pollution control. So sterility of process conditions is not thought of for digestion. Nevertheless, if the digested sludge is to be considered as an animal feed some control of pathogenic organisms is desirable, so this aspect of sterilisation or partial sterilisation will be considered later.

The Batch Culture

A digestion, being in effect a large bacterial culture, can be run in two ways; as a batch culture or as a continuous culture.

The batch culture is the type of culture carried out every day in the microbiology laboratory where tubes or flasks of medium are inoculated with a small amount of a previous culture and the medium is then incubated at an appropriate temperature for a suitable time. This suitable time is decided on not only by the growth rate of the bacteria but also by the use to which the culture is to be put, for example whether a large number of viable cells are to be obtained for inoculating another culture or whether a large amount of some end-product is required. This differentiation in incubation time is needed because of the growth characteristics of a batch culture.

The growth of a batch culture has been previously outlined and it follows the course shown in Fig. 3.1. After inoculation there is usually a 'lag period' at (a). This period varies in extent and may be more apparent than real. It is real in that if, for instance, the bacterial inoculum has been transferred from a different medium, or allowed to cool, or the new medium is not exactly correct for the bacteria, the bacteria may take some time to adjust or readjust enzyme systems to new conditions, or to adjust the medium to suitable conditions. How long this takes depends to some extent on inoculum size. On the other hand, the lag may only be apparent in that the detection system used cannot pick up the initial very small absolute change in bacterial numbers or products caused by growth of a small inoculum.

However, whether or not there is a true lag the bacteria will start growing and dividing as previously explained and this growth can be described mathematically in two ways.

Between time t_0 and t there are g generations of bacteria and so if the number of bacteria at time t_0 is n_0 and the number at time t is n_t then

$$n_t = n_0 2^g$$

Taking logarithms, then $\log n_t = \log n_0 + g \log 2$
or

$$\frac{\log n_t - \log n_0}{\log 2} = g \qquad (1)$$

If T is the mean generation time (the time for one bacterium to become two) then $t/T = g$ and substituting in eqn (1)

$$\frac{\log n_t - \log n_0}{t} = \frac{\log 2}{T}$$

Since n_0 and n_t may be measured, as number of bacteria, or as weights of bacteria, then T may be calculated.

Another way of expressing the growth of bacteria is by using the specific growth rate, usually designated as μ, where the rate of increase of a bacterial population is given by the expression

$$\frac{dx}{dt} = \mu x \tag{2}$$

where x is the number or weight (concentration) of bacteria and μ is the rate of increase per unit of organism concentration.

Integration of this gives

$$x_t = x_0 e^{\mu t}$$

or

$$\ln \frac{x_t}{x_0} = \mu t$$

The mean generation time is obtained by putting

$$x_t = 2x_0$$

when

$$\ln 2 = \mu T$$

or

$$T = \frac{\ln 2}{\mu} = \frac{0 \cdot 693}{\mu}$$

This type of expression was introduced by Monod (1942). The growth rate μ, all other things being equal, is controlled by substrate concentration and an expression generally used for the effects of substrate concentration on growth rate is

$$\mu = \frac{\mu_m S}{K_s + S} \tag{3}$$

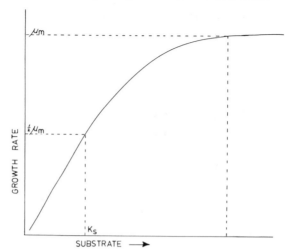

Fig. 4.1.   The dependence of bacterial growth rate on concentration of a substrate
(see text).

where μ_m is the maximum growth rate of the bacteria set by their ability to
take up and utilise the substrate, S is the substrate concentration and K_s is
the substrate concentration at which half maximum growth rate is
obtained.

If a bacterium is growing under conditions in which all factors except one
are in excess; let us say nitrogen source, growth factors and salts are all in
excess but the energy-producing substrate, perhaps a sugar, is in limited
amount, then the growth rate of the bacterium will depend on the amount of
sugar present, within a limit. When the sugar is in low concentration the
growth rate of the bacteria will be low as sugar molecules are, in effect, 'few
and far between' and the bacterium can take up these molecules only at a
low rate. As the concentration of sugar increases there comes a point where
the bacterium is 'saturated' with sugar molecules and is taking them up and
using them at its maximum rate and any further increase in sugar
concentration cannot increase the rate.

The dependence of growth rate on substrate concentration is shown
graphically in Fig. 4.1, where K_s and μ_m are indicated. The concentration of
sugar, or nitrogen source, or other factors such as magnesium or
phosphate, which limit the growth rates of bacteria are usually very low and
in most laboratory and factory batch cultures the amount of substrate
provided is well above this level. So in the batch culture described by Fig. 3.1
growth over the period (b) will be at maximum and constant rate described

by eqn (2) where $\mu = \mu_m$. However, as the substrate is used up there will come a time when its concentration falls into the region to the left of the curve in Fig. 3.1 and the bacterial growth rate will fall and finally become zero. However, it should be noted that although initially the substrate concentration is above that at which growth rate is controlled the substrate concentration can still influence the mass (or number) of bacteria obtained. Doubling the sugar concentration will (if nitrogen source and growth factors are still in excess) double the time for which the bacteria can grow and double the amount of bacteria produced, assuming that other factors, such as change in pH, do not affect the growth. This is expressed in another equation

$$Y = \frac{x}{Sub} \qquad (4)$$

where x is the weight of bacteria formed and *Sub* is the amount of substrate used, and Y is the yield factor or coefficient.

If a series of cultures is set up with different amounts of sugar and the rest of the medium the same in each case, then there could be a sugar concentration above which any increase makes no difference to the amount of bacteria obtained. The sugar has then become the excess substrate and concentration of some other substance, e.g. nitrogen source, is limiting the extent of growth.

These equations have formed the basis of many mathematical models for bacterial growth and for modelling of digester systems, but there have been very many extensions of these basic equations to account for the non-ideal behaviour of bacteria and to try to fit the equations to the growth of mixed cultures with a number of different stages in the culture such as is exemplified by the anaerobic digester.

In general the growth of aerobic bacteria, which have been used in many experiments, tends to fit these ideal curves best, provided aeration is sufficient. The growth of anaerobic bacteria with acidic and other fermentation products tends, in practice, to be more complicated. However, with all bacteria there are deviations from the ideal. These deviations are concerned with growth at low rates.

The equations (1) and (2) assume that all the bacteria in a culture divide and none die. This is almost true for many bacteria, particularly when the growth rate is high; the fraction of the bacteria which die and cease to divide is so small that any deviation from the ideal is impossible to detect.

However, when bacteria are growing at low rates the death rate can be considerable and so the behaviour deviates from the ideal. If the conditions in a culture, such as amount of substrate, are such as to induce a very low growth rate then the bacteria may not attain even the low growth rate predicted from extrapolations from higher substrate concentrations and the culture may effectively die out. High death rate of bacteria at low growth rates has been demonstrated experimentally in continuous cultures and these cultures deviate from the ideal.

In a mixed culture with particulate substrates and an input of live and dead bacteria as is a digester fed with faecal wastes, it is experimentally impossible to determine whether the low overall growth rate leads to a high death rate of bacteria.

The other deviation from ideal behaviour comes in the factor of maintenance requirements and uncoupled fermentations. The ideal bacterium uses all substrate to provide energy for growth processes, and in some cases movement, whatever the rate of growth of the bacterium or conditions in the medium, and it forms the same amount of cells per unit of substrate used at all growth rates. This is implied to some extent in eqn (4) and it is assumed to be correct later when the simplest models of continuous cultures are discussed. However, bacteria deviate from this constant yield factor, and in the case of some of the anaerobic bacteria such as are found in digesters the deviation is quite marked. Under some conditions, for instance if a nitrogen source is limiting extent of growth, and energy source (sugar, say) is in excess, then the bacteria will stop growing when the nitrogen source is used up, but they can continue to ferment the sugar, the energy produced being dissipated as heat. Other factors such as lack of a vitamin or a particular salt could stop growth with the same result. Similarly a suspension of viable bacteria transferred to a solution of a sugar and salts, with no nitrogen source, can ferment the sugar in the absence of growth. Such 'uncoupled' fermentation if it occurred in digesters could be beneficial, as it would result in degradation of a waste substrate and so decrease its polluting power, without increase in numbers of bacteria to add to the weight of digested sludge solids, but it would make the behaviour of the digestion deviate from theory.

The other way in which bacteria can use energy without growth is for maintenance. Bacterial cells, like animal bodies, are not in an unvarying state once formed. The constituents of the cell are constantly broken down and remade. This requires energy as any energy produced in breakdown of a constituent cannot be used for the synthetic processes. This use of energy is a 'maintenance' requirement. The use of energy for maintenance is more apparent at low growth rates than at high and so the yield of bacteria per unit of substrate used falls as the growth rate decreases. In some cases the yield seems to fall very rapidly as growth rate decreases from near maximum, in other cases the effect seems to be apparent only at growth rates a small fraction of the maximum. Again this phenomenon, which must occur in digesters, can only add to the efficiency of the digestion as a pollution control in that substrate will be used without adding to the weight of bacteria in the digested sludge.

So, although overall the time course of bacterial growth and gas production in a batch digester will follow the course of Fig. 3.1, this overall activity will be made up of growth curves of different bacteria using different substrates and producing the different products of the reactions previously discussed (see Fig. 3.2). Although in a faecal waste there are the bacteria necessary for digestion these are in various numbers and in different numbers from those of a balanced digester population. Thus even if the bacteria follow the ideal growth curve, those in largest number initially will more quickly attain a dense population and may overgrow others, or they may produce acids before the methanogenic bacteria, which are slow growing as discussed before and in small numbers in the excreta, can grow sufficiently to use the acids, so the digestion will go acid, stopping further growth. Some bacteria, for various reasons, may have so little substrate that their growth is slow, so they have a high death rate and never attain a proper population. To try to ensure that all the bacteria get a chance to grow in the proper proportions it is usual to give an inoculum of digesting sludge. This is generally done by leaving some of the digested sludge in the digester when it is emptied and refilled. But the inoculum bacteria must be still active and not too far into portion (c) of the growth curve or death through lack of substrate may have altered the balance of the population. Of course, with digester feedstocks that contain few bacteria an inoculum must always be given to start the digestion. The interdependence of the bacteria for substrates and growth factors thus adds complications and the system may not attain a proper balance and inadequate digestion will result. In addition, there are the practical problems of down-time and dealing with a constant flow of feedstock or the necessity for a continuous supply of gas.

The types of continuous culture that can overcome these latter problems are discussed next.

Continuous Cultures
The tubular fermenter: The tubular fermenter or digester is, as its name implies, a horizontal tube which is fed with substrate at one end while from the other end digested material is removed. The ideal tubular digester is

shown in Fig. 4.2a. There is no mixing and each slug of input passes along
the tube as a discreet body. In effect this is a batch culture in which the
reactions take place with a time and a space factor. The growth of the
bacteria and the production of gas follows the graph of Fig. 3.1, but each
segment of the curve of Fig. 3.1 is represented by a slug of material in the
digester, the left of the tube being part (a) on the curve, and the right part
(b)–(c). The length and volume of the tube must be related to the volume of

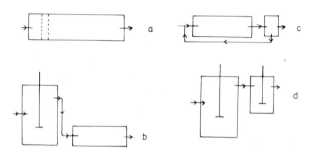

FIG. 4.2.   (a) A tubular digester. The arrows show input and output and the dotted
lines the theoretical slug of input passing along the tube. (b) A tubular digester fed
from a single-stage, stirred-tank digester. (c) A tubular digester with feedback of
bacteria separated from the output flow. (d) A two-stage stirred-tank digester. The
two tanks are of different sizes giving different detention times for the same flow rate.

input such that the time t in Fig. 3.1 is the time taken for a slug of input to
pass through the tube. The lag phase (a) is usually regarded as negligible.
   An inoculum of bacteria is needed at the input end of the tube just as it is
needed at time zero in a batch culture. If the input is animal excreta then, as
in the batch culture, the correct bacterial population could develop as the
excreta passed down the tube. But, again as in batch culture, to ensure that
the correct reactions take place it is better to inoculate with growing
bacteria from an established digestion. However, consideration of the
theoretical tubular digester shows that an inoculum mixed with one slug of
input will travel with that slug and that the feedstock before and behind will
remain uninoculated. So a continuous inoculation is required. In practice
this can be obtained in two ways. A tubular fermenter can be linked to the
output of a stirred-tank fermenter (see next section) as in Fig. 4.2b. This
system then forms one type of two-stage fermenter (see later), where the
substrate for the tubular fermenter is the residual substrate from the stirred-
tank fermenter.

At the simplest level, the reactions in an inoculated tubular fermenter can be considered as follows:

If t is the detention time of a slug of substrate in the tubular fermenter and x_1 the concentration of bacteria entering the fermenter then growth of the bacteria will increase the concentration by Δx where $\Delta x = x_1 e^{\mu t} - x_1$. So $x_1 + \Delta x$, the total concentration of cells at the end of the tube (x_t) is $x_1 e^{\mu t}$ as in a batch fermenter, where μ is the growth rate of the bacteria.

Then

$$t = \frac{\ln x_t - \ln x_1}{\mu}$$

and the dilution rate (see later) of the tubular fermenter (D) is given by

$$\frac{1}{t} \quad \text{or} \quad \frac{\mu}{(\ln x_t - \ln x_1)}$$

$$V = \frac{F}{D} = \frac{F(\ln x_t - \ln x_1)}{\mu}$$

The decrease in substrate (ΔS) is

$$\frac{\Delta x}{Y} = \frac{x_1 e^{\mu t} - x_1}{Y} = \frac{x_t - x_1}{Y}$$

This treatment holds only if μ is constant over time t, as was explained for the batch culture, so as in that case the substrate level must be above the growth limiting rate $(S \gg K_s)$ or t must be so short that little change in substrate concentration takes place.

The method of inoculating the incoming waste when the tubular fermenter is by itself is to have a feedback of bacteria from the end of the fermenter. Feedback is considered again later, but essentially either a part of the outflow is diverted and pumped back to mix with the incoming feedstock, or, more usually, the bacteria from part of the output are removed and fed back as a concentrated suspension in order to avoid diluting the feedstock with large volumes of spent medium. The feedback tubular fermenter is shown diagrammatically in Fig. 4.2c.

Expanding the treatment above the following is obtained.

If the flow of feedstock is F then the flow through the fermenter will be F_s where $F_s = F + aF_s$. (a is the fraction of the flow returned.) If the bacteria in this return flow are concentrated from the outflowing liquid by a factor c then the returned bacteria will be acx_R where x_R is the concentration of bacteria in the outflow from the fermenter.

The inoculum of bacteria is thus $ac x_R$ and this will increase to x_R where $x_R = x_R ac e^{\mu t}$.

So

$$\ln x_R - (\ln x_R + \ln ac) = \mu t$$

$$\ln x_R - \ln x_R - \ln ac = \mu t$$

$$- \ln ac = \mu t$$

and as before

$$t = \frac{V}{F_s}$$

so

$$- \ln ac = \mu \frac{V}{F_s}$$

$$\ln \left(\frac{1}{ac} \right) = \mu \frac{V}{F_s}$$

Since

$$F_s = \frac{F}{(1 - a)}$$

then

$$V = \frac{F}{(1 - a)\mu} \ln \left(\frac{1}{ac} \right)$$

The decrease in substrate is equal to the increase in bacterial concentration divided by Y.

As explained before, this latter model would describe growth limited by a sugar substrate where K_s is very small compared with the usual amount of substrate used in fermentations and so bacterial growth will continue at maximum rate until 95 % or more of the sugar is used up. The residual sugar returned in the feedback liquid will be negligible compared with that in the feedstock.

However, in the case of the stirred-tank fermenter linked to a tubular fermenter the conditions in the stirred tank will be such that the concentration of sugar in the effluent will have been brought to growth-rate-limiting values, as shown in the next section. As this sugar is used up the bacteria in the tubular fermenter will grow at declining rates as they pass along the tube and the simple treatment will not hold. Nevertheless the residual substrate from the tubular fermenter should be lower than that from the first stage, but whether such a system would be worthwhile in practice depends on factors such as those discussed for the two-stirred-tank fermenters below.

A more extensive treatment of the tubular fermenter is to consider a number of stirred-tank fermenters in series (two tanks are considered later) and extend this to a limit of an infinite number of tanks of total volume equal to that of the tubular fermenter. The paper by Powell & Lowe (1964) should be referred to for this treatment.

The theory of tubular fermenters considered growth of the bacteria as starting immediately the feedstock plus inoculum entered the tube. In the tubular digesters described later no inoculum, as such, is used, so none of the models will apply. The bacterial population develops from the bacteria in the animal waste used as feedstock, so if the digesters are behaving as a true tubular fermenter the length of the tube will have to be such as to contain the lag phase while the correct balance of organisms develops and then the rest of the tube could, in theory, be considered as an inoculated culture.

The models of tubular fermenters such as those above are based on one bacterial species and one substrate. As explained in the case of the batch culture, a digestion is much more complex, with at least two reactions occurring (fermentation and methanogenesis) and more usually three, as a hydrolysis of polymers generally precedes the fermentation. This first reaction can be slow and rate limiting as shown later. Since the reaction rate is slow, long detention times are required in a stirred-tank digester and this means that the concentration in the effluent of the volatile fatty acids which could form the substrate for a second, tubular, digester is low and below the K_s value. Although these concentrations could be lowered in a second digester it would probably, as shown later, only be worthwhile adding this in very special circumstances if at all.

But whether the fermenter is to be added to a stirred-tank, or whether it is to be a separate feedback system, the ideal tubular fermenter is almost impossible to attain in practice. Wall friction impedes the flow of, and tends to mix, the liquid. If it is heated, convection currents can again mix the liquid. In the case of digesters the gas formation will mix the 'slugs' of sludge theoretically passing along the tube, and, indeed, to keep solids suspended some mechanical mixing of the sludge may have to be used. The 'tubular' digester will in practice tend to become more akin to a long stirred-tank digester with a general flow from one end to the other.

The stirred-tank single-stage digester: The simplest form of continuous culture is a tank with inflow of medium and outflow of bacteria, metabolic products and depleted medium (Fig. 4.2b). This system is the 'completely-mixed, single-stage culture', or the 'stirred-tank' culture, ('mixed' is used in this context in the sense of stirred) and it is the most usual form of continuous culture used in the laboratory and industry and is the form taken by most anaerobic digesters. Its other name is the 'chemostat' culture. Digesters, though, deviate from the theoretically ideal conditions. The digester deviates from theory because it is a complex bacterial culture with complex actions and interactions and also because the substrates are often solids or polymers. However, the theory of continuous culture will be discussed because mathematical modelling is based on this theory either by assuming the culture is actually simple, or by endeavouring to apply the theory to each stage of the digestion process, an exercise which is now possible with the aid of computers.

The stirred-tank, continuous culture is based on the growth of one type of bacterium on one limiting substrate as previously described. This substrate then controls the rate of growth and concentration of the bacteria in the culture. The culture is contained in a tank and is completely homogeneous with respect to bacteria and medium constituents. The fresh medium is added as a continuous stream and immediately mixed into the tank contents and as fresh medium is added the bacteria and spent medium flow out at the same rate. There is also no adherence of bacteria to the walls of the tank. Obviously mechanical aspects make such a system impossible to attain in practice, but it can be attained in the laboratory sufficiently closely to show that the theoretical treatment is essentially correct and deviations occur because of non-ideal behaviour of bacteria as mentioned in connection with batch cultures.

The following theoretical treatment of continuous culture was originally put forward by Herbert *et al.* (1956) and is based on the Monod description of bacterial growth previously mentioned.

The Dilution Rate (D) of the culture is the inverse of the detention time of medium and bacteria in the system and is F/V where F is the flow rate of medium through the culture and V is the volume of the culture. D has dimensions of reciprocal time (e.g. 1/h) and F and V are in the same volume units.

The bacteria are growing in the culture at a rate defined by eqn (2) and equal to μx. They are also flowing out of the culture at a rate Dx.

So the rate of increase (dx/dt) in bacterial concentration is growth minus output which is

$$\mu x - Dx \qquad\qquad (5)$$

The rate of change of the limiting substrate concentration (dS/dt) in the culture is, input minus output minus consumption by bacteria.

The input rate is DS_R, where S_R is the substrate concentration in the incoming medium.

The output rate is DS, where S is the substrate concentration in the outflowing used medium. The rate of consumption by the bacteria is

$$\frac{\text{growth}}{\text{yield factor}} = \frac{\mu x}{Y}$$

Y is here assumed to be a constant for all values of μ, as previously mentioned.

So the rate of change in substrate concentration

$$\frac{dS}{dt} = DS_R - DS - \frac{\mu x}{Y} \tag{6}$$

The growth rate is related to the substrate concentration by eqn (3),

$$\mu = \mu_m \frac{S}{K_s + S}$$

so substituting in eqn (5) gives

$$\frac{dx}{dt} = x\mu_m \left(\frac{S}{K_s + S}\right) - Dx \tag{7}$$

and eqn (6) gives

$$\frac{ds}{dt} = DS_R - DS - \frac{x}{Y}\mu_m \left(\frac{S}{K_s + S}\right) \tag{8}$$

If the culture is in a 'steady-state', i.e. neither the concentration of bacteria nor substrate is changing, then

$$\frac{dx}{dt} = \frac{dS}{dt} = 0$$

Equating dx/dt and dS/dt to zero gives steady-state values of x and S, $\bar{x}$ and $\bar{S}$, of

$$\bar{S} = \frac{DK_s}{(\mu_m - D)} \tag{9}$$

$$\bar{x} = YS_R - K_s \left(\frac{D}{\mu_m - D}\right) \tag{10}$$

The steady-state concentrations of bacteria and substrate in the culture depend on S_R and D.

At the steady state, since from eqn (5)

$$\frac{dx}{dt} = 0 = x(\mu - D)$$

then $\mu = D$. That is, the bacteria are growing at the rate at which the medium is passing through the system. The critical or 'washout' dilution rate D_c is defined by

$$D_c = \mu_m \left(\frac{S_R}{K_s + S_R} \right) \tag{11}$$

that is, the highest possible value of μ when S is the highest possible value, S_R. But at this value of S, x is zero as no substrate would be used for growth, so as D approaches this critical value x will fall rapidly and the culture will 'washout'.

It can also be shown mathematically that the output of a continuous culture in terms of cell mass (or a product related to cell mass) is greater than that of a batch culture over the cycle time (down time, lag phase, growth phase) of the batch culture.

These equations lead to a system in which the steady-state values of x are related to D as in Fig. 4.3 line (a).

The growth of some bacteria follows quite closely this theoretical graph as Herbert et al. (1956) demonstrated in their original tests of the theory. But more extensive experience with continuous cultures has since shown that in many cases deviation from the theoretical occurs. The deviation most likely to occur in digesters is where the bacteria use substrate for maintenance and Y decreases as D decreases. The values of x then decrease with decrease in D as in curve (b), Fig. 4.3.

A number of mathematical treatments of maintenance have been proposed, the most used one is that of Pirt (1965) who assumed a constant maintenance coefficient represented as follows:

$$\frac{1}{Y} = \frac{m}{\mu} + \frac{1}{Y_G} \tag{12}$$

where m is a maintenance coefficient such that the rate of substrate utilisation for maintenance is $-[(dS)/(dt)M] = mx$. Y_G is the growth yield corrected for maintenance requirements. If m and Y_G are constants then a plot of $1/Y$ against $1/\mu$ will be a straight line. Values for μ can be obtained by

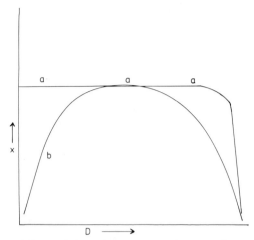

FIG. 4.3. Bacterial growth in a chemostat continuous-culture with one growth-limiting substrate. Curve a–a is the theoretical curve. Curve b is more like the results usually obtained in practice because of maintenance and other effects (see text). x, bacterial concentration; D, dilution rate.

equating this with steady-state values of D in continuous cultures and so m and $1/Y_G$ and hence Y_G may be found.

Some bacterial cultures seem to follow this model, or at least approximate closely to it, but there are deviations, and it is still not decided whether the maintenance requirement is constant or whether this varies with growth rate D.

The other deviation from the simple model is concerned with the death rate of the bacteria. Powell (1965), for instance, has considered this mathematically as an extension of the model given above, and it can be shown that if some bacteria die then the growth rate of the others must be greater than D for a steady-state culture to be possible. Tempest *et al.* (1967) demonstrated experimentally that this was so. In cultures of an aerobic bacterium at very low dilution rates (equivalent to a detention time of 100–300 hours) they found that the bacterial viability fell to only about 60%, that is, nearly half the bacteria were dead. As Powell showed, in this kind of situation while the doubling time of the whole culture is, as shown before, $\ln 2/D$, the doubling time of the viable cells is $(\ln 2 + \ln \alpha)/D$ where α is an 'index of viability' equal to $(v + 1)/2$ where v is the fraction of cells which is viable. This kind of finding implies that in a completely-mixed digester of low turnover rate many of the cells will be dead and that the digester culture cannot be treated as a simple chemostat culture.

The basic continuous-culture model has, however, formed the basis for a number of mathematical models of anaerobic digesters. But it should be noted that this theory applies only to a completely-stirred culture with medium inflow and outflow in a uniform and continuous flow, with the bacteria dependent on one limiting substrate in the medium. In a digester the situation is obviously more complex, with a number of reactions and a number of bacteria. To apply the previous models of bacterial growth to a digester means simplifying the system. One reaction, and this has usually been taken as methanogenesis, is taken as controlling the rate of the whole system, so it can then be considered as in effect a continuous culture of one type of bacterium. Since the model system is simplified, the results that can be obtained from the model are simple ones.

The various actions in a digester and their possible effects on the theoretical treatment of continuous-culture growth described above have been considered previously in reviews (e.g. by Kirsch & Sykes, 1971; Hobson *et al.*, 1974) and they will be mentioned here only briefly to indicate some of the difficulties in modelling digesters.

The theory assumes that the medium entering the culture vessel is sterile. This applies to the laboratory or industrial pure culture, but digesters will never be run with a sterile feedstock as explained before, and (also as previously described) faecal wastes contain very large numbers of bacteria. These range from ones which cannot grow in the digester to others which can metabolise and grow to some extent but which without continued inoculation would washout, and those which form the important fermentative and methanogenic population of the digester. The continued introduction of these latter bacteria is a deviation from theory, and can be in some sense equivalent to wall-growth in a pure culture system. The theory assumes all bacteria are free and dispersed in the medium. If bacteria grow on the wall of the culture vessel ('wall-growth') then they can break free at intervals and apparent steady states can be obtained at values of D above D_c. Apparent Y values are also affected.

Theory also assumes complete stirring and mixing of the culture. In a digester this is often probably not attained. In addition, many of the digester bacteria will be growing not dispersed in the medium but attached to particles of vegetable or animal debris which are the substrates. Conditions for growth of the bacteria can then vary at different microhabitats within the digester. If the digester is not properly mixed and layering occurs then conditions for growth of the bacteria will vary over large volumes of the contents and in effect there could be a series of cultures growing under different conditions at different heights in the digester tank.

The closest approximation to the ideal continuous culture will be given by a digestion of a simple sugar substrate (say molasses which is mainly sucrose, glucose and fructose) in solution. Such sugars are fermented rapidly and almost completely by bacteria and so the system may, in effect, be considered as one in which a substrate of acids, hydrogen and carbon dioxide is provided for the methanogenic bacteria at a rate and concentration defined by the input rate of the sugar solution and the concentration of the sugar. The residual acid level (S above) and substrate conversion efficiency, the maximum values of D (or minimum detention time), and the gas production at different detention times may then be calculated if various constants defining the growth of the methanogenic bacteria are known. If, however, the substrate is a polymeric carbohydrate such as cellulose, or a lipid for instance, then the reactions can be more complex. The supply of sugar for fermentation and so the supply of acids for the methanogenic bacteria may be governed by the rate of hydrolysis of the polymeric carbohydrate. If the hydrolysis is slow then the maximum dilution rate at which the digestion can be run will be governed by the time required for appreciable hydrolysis of the carbohydrate and this may be less than the maximum dilution rate for the methanogenic bacteria, and the amount of gas produced will depend not on the amount of carbohydrate entering the system and the residual acid left by the methanogenic bacteria at any dilution rate, but on the amount of carbohydrate hydrolysed to sugars at any particular dilution rate. The effects of slow hydrolysis of cellulose on critical dilution rates (detention times) have already been discussed.

Using continuous cultures of one bacterium it has also been found that production of enzymes can be a maximum at one particular growth rate, or even that production has peaks at two different growth rates. Hobson & Summers (1966) and Henderson *et al.* (1969) have shown that lipase, amylase and protease activities of bacteria similar to those found in anaerobic digesters can behave in this way and other hydrolytic enzymes may be similar. If digestion primarily depended on hydrolysis of a polymer then the maximum rate of hydrolysis of this, if it depended on the specific enzyme activity of the bacterial cells, would occur at a particular detention time. This could be a factor making for deviation of an actual system from the model.

Various workers have attempted to take into account the different reactions in the usual digester system by considering each step in the process and applying differential equations to these. Some authors have also considered the inhibitory effects of accumulation of acids and the other factors mentioned in Chapter 3. Such models have then been applied to predictions of digester behaviour with various feedstocks or to behaviour of different types of digesters with one feedstock. In a recent paper Hill & Nordstedt (1980) used differential equations in computer models of agricultural-waste digestions. The computer was able to simulate non-steady-state systems such as the failure of a digester on overloading. The various rate constants have to be obtained from experiments, however. These figures can be obtained from experiments on different feedstocks, but the figures for breakdown of a particular substrate may be affected by the presence of other materials in the feed. For example, Lawrence (1971) found that there were different values for coefficients of long-chain fatty acid breakdown in digestions where pure acids, only, had been the feedstock and in digesters where lipid material in sewage sludge was being digested. In addition, the substrate called by a particular name in one feedstock may not be the same as the same-named material in another; this has been mentioned already for cellulose for instance. In determining the various constants for digestion of a particular feedstock so many experiments may have to be done that the necessary information for proper digestion of that feedstock is obtained and a model can give no further useful information.

However, if the information available on digestion of a particular feedstock does fit a model of the system based on theories of bacterial growth and on the reactions taking place in the system, then it shows that these theories must be correct or approximately correct. By taking information gained with one or a number of feedstocks or feedstock components it may be possible to predict at least approximately the conditions for digestion of another feedstock of known composition. And models may be used, even if not absolutely correct, to determine the relative merits of different digester systems, say a single stirred-tank and a two-phase system, for a particular feedstock or a particular type of feedstock.

For instance, Hobson & McDonald (1980) showed that the concentrations of residual acids in a piggery-waste digester run at different dilution rates (detention times) at 35 °C were the same as those in a continuous culture of bacteria converting acetic acid to methane run at the same dilution rates. The concentration of acid at any dilution rate was described by eqn (9) above and the bacteria converting acids to methane behaved in the manner predicted by the continuous-culture theory above with a K_s of 800–900 mg acid/litre and a μ_m of 0·4/day. However, it was shown that the digester could only run at dilution rates near this μ_m because the feedstock contained a high concentration of volatile acids. In effect what was being modelled was a stirred-tank continuous culture of bacteria converting acids to methane running separately from another culture in which the solid substrates were being converted to acids. This latter would add to the S_R value of the methanogenic culture an amount of acids varying with digester dilution rate, but as shown above S_R does not affect the residual substrate concentration. Further (unpublished) calculations showed that the breakdown of solids did not fit the simple continuous-culture theory and it seemed that it probably consisted of a number of bacterial reactions of different kinetics. This could be caused by breakdown of cellulose, hemicellulose, fats, etc., by different bacteria.

Some digesters are run, or have been run, on the basis of an overall detention time of, say 30 days, but with feeding and withdrawal of large amounts of feedstock and digested sludge at intervals of some days. Such running deviates from the running of the ideal continuous culture and the digester becomes more of a 'fed-batch'- than a 'continuous'-culture. The theory and modelling of the fed-batch culture have been considered by some workers and, for instance, Pirt (1974) gives a comprehensive treatment. As with the continuous-culture theory this deals with one substrate and one bacterium. It will not be considered further here as present practice is to run the stirred-tank digester as near as possible to the theoretical continuous culture.

The two-stage digester: It was previously mentioned that the time for which a batch culture is run depends on whether the maximum amount of viable cells is needed or whether some product formed during active growth of the bacteria or some product formed by the cells in the resting stage after active growth has ceased is required. As an example of the latter some microbial products are formed by the organisms metabolising excess carbohydrate present after growth has been limited by the nitrogen content of the medium.

Since a stirred-tank chemostat culture is designed, and indeed works, on the basis of keeping the organisms growing at a constant rate, it is obviously unsuitable for production of a substance formed during a non-growth stage of the microbial life cycle. The chemostat continuous culture can be adapted to such a product formation by using a two-stage system where the micro-organisms are grown in one stirred-tank continuous culture under, say, nitrogen limitation and the culture then allowed to flow into a second tank where the cells can metabolise the excess carbohydrate in the medium under resting conditions. The detention time in the second tank can be altered from that in the first tank for any particular flow rate of medium by making the second tank a different size from the first. By addition of acid or alkali the pH in the second tank may be adjusted from that of the first, or the temperature may be altered, or aeration rate of an aerobic system might be adjusted. In addition, another substrate for bacterial metabolism might be added.

In the second application of the two-stage system, essentially the same reactions go on but a combination of two tanks may give a better utilisation of substrate than one tank.

The theory of continuous culture described above can be extended to two cultures in series as in Fig. 4.2d.

The first tank operates at steady-state conditions (dilution rate $= D_1$) as described previously and the concentration of residual substrate (S_1) leaving the first tank and entering the second is defined by eqn (9). The concentration of bacteria (x_1) entering the second tank is defined by eqn (10).

In the second tank the concentration of bacteria will be x_2 and the dilution rate D_2. So, the rate of increase of bacteria

$$\frac{dx_2}{dt} = \text{growth rate} + \text{input rate} - \text{output rate}$$

$$= \mu_2 x_2 + D_2 x_1 - D_2 x_2$$

At steady state

$$\frac{dx_2}{dt} = 0$$

so

$$\mu_2 \bar{x}_2 = D_2 \bar{x}_2 - D_2 \bar{x}_1$$

$$\mu_2 = D_2 - D_2 \frac{\bar{x}_1}{\bar{x}_2} \tag{13}$$

$$\bar{x}_2 (D_2 - \mu_2) = D_2 \bar{x}_1$$

and

$$\bar{x}_2 = \frac{D_2 \bar{x}_1}{(D_2 - \mu_2)} \tag{14}$$

The substrate balance for the second stage is given by:

$$\text{rate of increase} = \text{input rate} - \text{consumption} - \text{outflow}$$

or

$$\frac{dS_2}{dt} = D_2 S_1 - \frac{\mu_2 \bar{x}_2}{Y} - D_2 S_2$$

At steady state

$$\frac{dS_2}{dt} = 0$$

So

$$D_2 S_1 - \frac{\mu_2 \bar{x}_2}{Y} - D_s \bar{S}_2 = 0 \tag{15}$$

Substituting for μ_2 from eqn (13) and putting $\bar{x}_1 = Y(S_R - \bar{S}_1)$, eqn (15) reduces to

$$\bar{x}_2 = Y(S_R - \bar{S}_2) \tag{16}$$

Equation (13) shows that $\mu_2 < D_2$ and provided $\bar{x}_1$ and $\bar{x}_2$ are finite no washout will occur whatever the value of D_2. The first stage will, of course, washout when D_1 becomes greater than μ_m. Equation (16) shows that the concentration of bacteria in the second stage will equal that in a single-stage culture run at a dilution rate such that $S = \bar{S}_2$.

Putting $\mu_2 = \mu_m \bar{S}_2/(K_s + S_2)$ into eqn (15) and then $\bar{x}_2$ equal to $Y(S_R - \bar{S}_2)$ gives an equation in $\bar{S}_2$

$$\bar{S}_2^2(\mu_m - D_2) + \bar{S}_2(D_2\bar{S}_1 - \mu_m S_R - D_2 K_s) + D_2\bar{S}_1 K_s = 0 \tag{17}$$

from which it follows that

$$\bar{S}_2 = $$
$$\frac{-(D_2\bar{S}_1 - \mu_m S_R - D_2 K_s) \pm [(D_2\bar{S}_1 - \mu_m S_R - D_2 K_s)^2 - 4D_2\bar{S}_1 K_s(\mu_m - D_2)]^{1/2}}{2(\mu_m - D_2)}$$

This can only be solved numerically, and as an example of the behaviour of a two-stage digestion predicted by this model the piggery-waste digestion used in the paper quoted above by Hobson & McDonald (1980) can be used. The substrate coming from the first stage is volatile acids which are further converted to methane in the second stage. If the first stage is run at different dilution rates (i.e. detention times), but at 35 °C, then the values of $\bar{S}_1$ can be obtained from the graph of residual substrate against detention time in the first stage. These values for detention times of 3, 4, 6, 10, 20, 30 days were 4300, 1500, 650, 350, 150 and 75 mg/litre, respectively. Since the reactions being modelled are the same in the two stages, the values for μ_m, K_s and S_R are those calculated for the first stage; i.e. 0·4/day, 900 mg/litre, and 5000 mg/litre.

Calculation of $\bar{S}_2$ shows that one of the roots of eqn (17) is absurd in

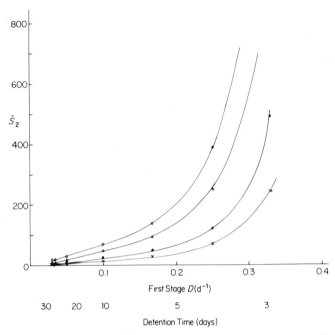

FIG. 4.4.   The theoretical steady-state residual acid concentrations $(\bar{S}_2)$ in the second stage of a two-stage digester with a piggery-waste feedstock to a first-stage of detention time (dilution rate) shown. Second stage: 10 day detention, × ; 6 day detention, ● ; 3 day detention, △; 2 day detention, ○.

each case, so taking the reasonable values a series of curves can be drawn up as in Fig. 4.4.

   These values show that the second stage can reduce the residual acids in the first stage effluent, and that a stable second stage could be run at a detention time (two days) which would cause washout of the first or a single stage. However, consideration of the single-stage piggery-waste digesters shows that a detention time of at least 10 days is required to get maximum breakdown of solids, so that residual acids would be low. Whether in such circumstances it would be worth the mechanical complexity and the additional energy input in stirrers, tank heaters, etc., of adding a second stage to further reduce the acids and obtain a small amount of additional gas, is doubtful. If a final effluent of high quality were wanted for pollution-control purposes, then a second stage might be a reasonable proposition.

   Since the breakdown of the solids is complex it is difficult or impossible to describe from the simple model the results of running a first stage at a low detention time, and a second stage at a low detention time, but one might not expect as high a breakdown of solids as in a single stage run at the optimum longer time.

In the normal sewage digestion process a type of two-stage system is used in that the digester overflow is led into large open tanks where it remains for a number of days or weeks at ambient temperature while the sludge solids settle and consolidate. During this time a slow bacterial action continues and some of the sludge material is gasified with a resultant decrease in residual pollutants and sludge solids. A similar improvement in quality of digested piggery waste takes place as the sludge is stored in a lagoon or tank before spreading (Mills, 1977). This type of second stage is not only run at a temperature different from that of the first stage but it is not run as a continuous culture. The sludge is allowed to build up in a tank and after settling, the supernatant liquid and the 'thickened' sludge are run off separately, or the tank is remixed and completely emptied after a storage period when the sludge is required for use as fertiliser. The kinetics of the process are then more akin to a batch culture in which the residual biodegradable material from the tank digester is the initial substrate.

The multi-stage digester has also been considered purely from the point of view of a chemical reactor. Clausen *et al.* (1979) in considering a model system for digestion of American energy crops used an equation from *Chemical reaction engineering* (Levenspiel, 1972) to deduce that four digesters in series would be better than one.

The model is as follows (but it deals only with tanks of equal volume):
If the rate of reaction in a stirred, continuous reactor is r then

$$r = \frac{(C_{s0} - C_s)}{\theta}$$

Where C_{s0} is influent substrate concentration, C_s effluent substrate concentration and θ is detention time in days. For the relatively few results of the small-scale tests of herbage digestion carried out by Clausen *et al.* the relationship of r to C_s was approximately linear, so $r = kC_s$ where k is a rate constant (1/days).

The single tank volume for a particular digestion can be determined and the volume of each of a series of equal-volume digesters (v) is calculated from

$$v = \left[\frac{1}{k} \left(\frac{C_{s0}}{C_s} \right) \frac{1}{N} - 1 \right] Q$$

where Q is the flow rate of input and through the system and N is the number of tanks.

Application of this model to some of the authors' results for single-stage farm-waste digestions suggests that if the theory is applicable a single tank could be replaced by two or four equal tanks of smaller total volume than the single tank. However, the added complexity and energy costs of heating, stirring and generally running a multi-tank system would have to be set against any decrease in overall detention time.

This question of added complexity of building and operating and energy inputs into the systems is one which will probably override most theoretical predictions of somewhat greater efficiency, particularly for the small and medium-scale digester.

However, there is an extension of the two-stage digester system which could be viable and have advantages in some cases, and this is described in the next section.

The two-phase digester: In the two-stage digester, bacteria of one kind grow in the first stage and then these same bacteria carry out another reaction under different conditions in the second stage. It is also possible to have two fermenters, in the first of which one type of bacterium carries out a reaction which provides a substrate for another type of bacterium growing under different conditions in a second fermenter. This is the principle of the 'two-phase' digester.

The digestion process consists essentially of two steps, a fermentation of sugars and breakdown of fats and proteins with production of acids, hydrogen and carbon dioxide, followed by conversion of these acids and gases to methane and carbon dioxide. The first stage may be extended by an initial hydrolysis of polysaccharides. These two sets of reactions may not have the same conditions for optimisation and so when they are taking place in one reaction vessel they may be running at some mean value suboptimal for each, or only one may be at optimum. If the two systems could be separated then each could run at optimum condition.

If digestion were a defined process running on a sterile feedstock and producing a high-value product, then it would be possible to inoculate a first digester with one set of bacteria, then sterilise in some way the effluent from this fermenter and pass it to a second fermenter which could be inoculated with the requisite second bacteria which would form the desired product. As it is, digestion is a process in which the bacterial population is largely undefined by cultural work (only limited isolations have been done with a few substrates) and the population develops naturally or by undefined inoculation from another digester. The feedstock is not sterile and in the case of faecal wastes it continuously inoculates the digester with the bacteria for both stages. The product is cheap and so no expensive primary or intermediate sterilisation can be used. So the two reactions cannot be separated by the growth of inoculated, separate bacterial populations. If the two reactions had very widely separated pH optima then the bacteria could be mutually excluded by running the two stages at these different pH values. However, previous consideration of the process suggested that the optima are not too far apart.

The other way in which two stages in a bacterial process might be separated is to run them at different dilution rates. If the μ_m values for the different groups of bacteria differ considerably, then one group might be excluded from a fermenter by running it at a dilution rate higher than μ_m for the other group. This is simple to do in practice as it rquires only two different-sized tanks as already explained and no elaborate equipment for adjusting pH between tanks, or for sterilising the outflow of one tank. And, in addition, it does not require a defined inoculation as if an inoculum of the complete population is given, the slower-growing part of the population will automatically be excluded from the tank with the high dilution rate.

However, there is one drawback to this and that is that with a system such as a sewage or animal excreta digester which is being continually inoculated, the process with the highest dilution rate must be the first or the slower-growing bacteria cannot be excluded. Thus, if the fermentation stage is the most rapid then this can be carried out in a digester in which the rapid turnover will exclude the methanogenic bacteria and these can then be grown in a slow-turnover second digester. The rate of the primary hydrolysis and fermentation step then is the crucial factor in determining whether a two-phase digester system based on differing retention times (dilution rates) is feasible, and this is still a matter for discussion and experiment.

Hamer & Borchardt (1969) carried out experiments with a dialysis fermenter system to determine the parameters for optimum operation of the two phases of the digestion process. Their apparatus could not be made for large-scale use as it would be too complicated and costly, but it was suitable for laboratory experiments. Two stirred fermenters were used and the contents of each were circulated by pumps through a dialysis unit and back to the fermenter. In this unit the liquids were separated by a vinyl–plastic membrane. This membrane was permeable to small molecules such as volatile fatty acids. If a solution of acids is placed on one side of such a membrane and water is on the other side, then the acid molecules will pass through into the water until they are in equal concentration on each side. If the water is flowing and the acids are removed from it by some means, then as the 'clean' water circulates back to the membrane it will pick up more acids and so the acids on one side of the membrane can be transferred entirely to a reaction on the other side. Thus the acids formed by sludge digestion in one fermenter could be transferred to the second fermenter where, because there were only acids as substrate, only the acid-utilising methanogenic bacteria could survive.

The fermenters were inoculated from a domestic sludge digester, but only one was fed with sewage sludge on a 10-day detention time basis. The system did not entirely separate the two phases as gas was formed in the first, fermentation-step, digester.

The experiments were detailed and complicated, but the principal results showed that the optimum oxidation–reduction potential (measured between platinum and calomel elecrodes) for hydrolysis and fermentation of sewage sludge was between -508 mV and -516 mV, and the optimum pH was 6·90–7·00. For methane production from the volatile acids the values were -520 to -527 mV and pH 7·05–7·20. They said that the hydrolysis and fermentation stage was rate-limiting in sewage-sludge fermentation and that the fermentation step would be operating at 75 % of full efficiency at the potential of -523 mV and pH 7·1 which was the mean optimum for methane production. However, the optima are so close that no complete separation of the two processes could be made purely on pH or oxidation–reduction potential even if these could be controlled.

Others have attempted phase separation by detention time. The experiments of Ghosh & Klass (1978) have already been mentioned. They considered two mesophilic stirred-tank digesters. Laboratory experiments were done with undefined, mixed cultures of bacteria developed from a digester population by inoculation into media containing only acetic acid or carbohydrates as substrates. (It might be noted that digester filtrate had to be added to the acetic acid medium to provide unknown growth factors for the methanogenic bacteria). Continuous and semicontinuous cultures were run to determine growth rates of the bacteria, K_s, T and Y values, using in models essentially the continuous-culture theory previously described. They concluded that a two-phase digestion could be carried out by using different detention times in the acidification and methanogenic stages. These authors found that the optimum pH range for the acidification of a domestic sewage sludge was 5·7 to 5·9. This is much lower than the Hamer & Borchardt (1969) results quoted previously. The sewage sludge used by Ghosh & Klass (1978) came from Chicago sewage works and consisted of 90 % activated (i.e. aerobic treatment plant) sludge and 10 % primary sludge; this seems a lower proportion of primary sludge than is usually found in digester feedstocks. It is not certain what proportion of activated sludge was in the sludge used by Hamer & Borchardt. Ghosh & Klass found the optimum pH for acid conversion to methane was 7·0 to 7·4. Again, the phases were not completely separated and some biogas was produced in the fermentation stage.

Cohen *et al.* (1979) used laboratory apparatus to investigate two-phase mesophilic digestion of glucose solution. They operated the acidification fermenter at pH 6·0 and the methanogenic fermenter at pH 7·8. The detention times were 10 hours and 100 hours respectively. Ghosh & Klass used acetic acid as the test substrate in their two-phase system as they said that the main acid formed in digesters is acetic. However, the theoretical reasons for the formation of acetic acid in conjunction with the methanogenic bacteria have already been discussed. Cohen *et al.* found in conformity with this theory that in their fermentation stage the products were hydrogen and carbon dioxide with mainly butyric acid (about 1400 parts) with acetic acid about 500, propionic 40, formic acid about 50 and some lactic acid. The acid solution then formed the feed for the methanogenic fermenter. The hydrogen gas represented a loss of about 12 % of the digestible feedstock. To run the acidification fermenter at pH 6, the pH had to be automatically controlled.

These results show that a two-phase digestion is possible using different detention times, but only if the feedstock is suitable in that the fermentation phase is rapid. If the fermentation stage is slow because of the nature of the substrate as previously pointed out, then methanogenic bacteria would grow in the first stage and no phase separation would be possible.

With the object of producing acetic acid, which can be converted chemically into liquid fuels, or producing a mixture of volatile fatty acids which can be used as a feedstock for growth of microbes for single-cell-protein (SCP) production, some experiments have been done to try to stop digestion at the acidification stage, or at least greatly decrease the conversion of acids to gas. Such experiments are outside the scope of this book, as it is principally concerned with digestion as a producer of gas, and so will be only briefly mentioned later. While there may in the future be a demand for acetic acid to replace the chemically-prepared acid, there would seem to be other, more economic, ways of producing SCP than from a partial digestion. However, the basis of these experiments is the same as that for phase-separation described above.

The single-stage culture with feedback and the tower digester: In the previous section reaction rates of the various stages of the digestion process were brought into the discussion, and it was shown that the growth rate of the methanogenic bacteria is such that a detention time of some four days is required to prevent washout of the bacteria from the digester. However, the growth rate of the fermentative bacteria is such that for simple substrates a detention time of only a few hours can be used for the acidification stage without washout of the bacteria. Many factory wastes consist of waters used for washing, blanching, boiling, etc., of vegetables and meats and contain only small amounts of material in suspension, most of the contents of the waste being dilute solutions of sugars and other relatively-easily biodegradable material. However, their main characteristic is that they are produced in large volume at high flow rates. To detain a flow of some hundreds of gallons a minute for a detention time of even four days would require a digester of impossibly large dimensions. And while the hourly or daily gas production from such waste-waters might be high, because of the low substrate concentrations the gas production per unit volume would be low. A large digester combined with a low volumetric gas production is not an economic process.

In such situations a digester with feedback is a possible solution. If a portion of the bacterial mass in the digester effluent is returned to the digester, then the continuous-culture theory previously described no longer applies. There is, in effect, a 'wall-growth' which, as previously explained, allows operation of the culture at above the washout dilution rate, because although bacteria are washing out before they can divide the culture is being continually reinoculated from the mass of bacteria growing on the wall. Wall-growth, however, is uncontrolled, but if a controlled amount of bacteria is returned to the digester then the detention time of the bacteria in the digester can be increased so that a steady-state growth can be obtained while the detention time of the liquid is less than that of the bacteria. In practice, detention times for the bacteria sufficient to allow growth of the methanogens (i.e. some days) can be obtained with a liquid detention time of only a few hours. Such feedback digesters have been referred to as 'contact' digesters.

Pirt & Kurowski (1970) have provided a comprehensive mathematical treatment of continuous fermenters with feedback. They considered three systems, all of which are relevant to digester practice, in addition to the simplest feedback. This latter, shown in Fig. 4.5a was originally considered theoretically by Herbert (1961). A portion of the microbial mass is sedimented out in the sedimenter and returned to the fermenter. In practice it would be difficult to separate off exactly the right amount of concentrated micro-organisms for return, so a more practical apparatus is shown in

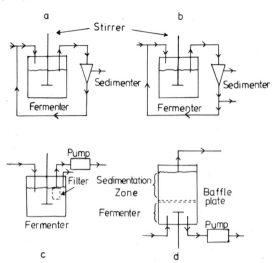

FIG. 4.5. Various forms of single-stage, continuous culture with feedback (after Pirt & Kurowski, 1970). Model (a) is a system with feedback of all sedimented bacteria; model (b) allows a proportion of the sedimented bacteria to be returned. Model (c) is equivalent to a digester with internal baffles or some other system for keeping part of the bacterial mass in the digester. Model (d) is on the same principle as the sludge-blanket or the fluidised-bed digester.

Fig. 4.5b, where the amount of concentrated microbial suspension can be adjusted by the junction pipe below the sedimenter where a portion of the sedimented micro-organism stream can be bled off to waste. Figure 4.5c shows a fermenter with a filter on the outlet which prevents microbes from leaving the fermenter with the used-medium effluent stream. Since a continuous build up of microbes in the fermenter would eventually lead to failure, some of the concentrated microbial suspension in the fermenter must be removed as well, as shown in the figure. This filter type of feedback system was the one used by Pirt & Kurowski to test their mathematical models. They used a laboratory-scale fermenter growing a yeast in a glucose medium. The final model is one in which the micro-organisms are allowed to settle in the fermenter (Fig. 4.5d) and the main used-medium flow is from the top of the fermenter while the concentration of bacteria in the fermenter is kept constant by removing sedimented bacteria from the bottom.

The theoretical treatment of Pirt & Kurowski will not be considered here except to say that it involved the 'Monod' principles of growth and relationship of growth rate to substrate, and the constant yield factor used in the previous treatment of continuous-culture growth. The authors found that the 'critical', or 'washout' dilution rate for the systems was not equal to μ_m, the maximum growth rate of the organisms in the medium (see earlier), but was μ_m divided by a factor less than one. The reciprocal of this factor expressed the increase in microbial concentration over that which could be obtained in a continuous culture without feedback and was called the 'feedback factor'. In the experimental filter-culture system μ_m for the yeast in batch or ordinary continuous cultures was 0·42/h, thus the continuous culture washed out at this dilution rate. However, with different feedback factors (in this case, since bacteria-free medium was extracted through the filter, equal to $1/C$ where C is the fraction of the medium flow taken out in the concentrated microbial stream and $(1 - C)$ the fraction taken out through the filter) the washout dilution rate could be extended to over 1·6/h.

The 'filter' feedback fermenter of Pirt & Kurowski corresponds with a digester in which the bacteria are retained in the digester by some form of separator integral with the digester overflow system. Such a separator cannot produce a bacteria-free effluent in practice as it would become blocked, so that the practical digester would not conform to the model mentioned in the past paragraph, but to the model also considered by Pirt & Kurowski in which micro-organisms are removed in the effluent stream passing through the filter.

The model in which a sedimenter returns a concentrated microbial suspension corresponds with the generally-used type of contact digester described later, while the 'sedimentation' model corresponds with the 'sludge-bed', tower digester.

The anaerobic filter and the fluidised-bed digester: These digesters are in many respects similar to the 'sedimentation' feedback fermenter described above. The object is again to obtain a long detention time for the bacteria with a short liquid detention time. In these cases the bacteria are retained in the digester by allowing them to attach to a solid support matrix, and in this they are like the aerobic 'trickling filter' used in sewage treatment. However, unlike the trickling filter the system is contained in an air-tight vessel and the liquid movement is from the bottom upwards (Fig. 4.6).

In the anaerobic filter, first studied on a laboratory scale by Young & McCarty (1968, 1969) the bacteria are retained on a matrix of relatively large-sized pieces of rock or other material which is packed immovably in the digester tank. The upflowing liquid passes through the spaces between the packing pieces and purified effluent runs over a weir or other outflow at the top. In the fluidised-bed digester the bacteria are retained by growth on very small (c. 1–2 mm) glass spheres. These spheres give a very much larger surface area for bacterial growth and contact with the liquid (and therefore

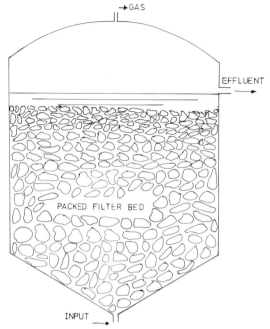

FIG. 4.6.   The anaerobic filter.

substrates) than is provided by an equal weight of large support pieces and as they are kept in suspension and movement by the upflowing stream of liquid, contact between bacteria and liquid is again enhanced. The fluidised-bed digester has been developed and tested on a small scale by, for instance, Jewell (1980) and some results will be considered later.

Obviously, in both systems a point will come when bacterial growth becomes so heavy that it will break away from the support matrix, and this loose growth will be removed in the effluent stream from the top of the digester or will settle to the bottom of the digester and have to be removed as a sludge from underneath the matrix. Such systems are suitable only for feedstocks containing dissolved substrates and very fine particulate material. Large particles would rapidly block a filter system and interfere with the suspension of a fluidised bed.

Young & McCarty (1969) considered the anaerobic filter in theory, and Shieh (1980) has recently produced a model for the fluidised-bed reactor.

Digesters for 'Solid' Feedstock

Some waste materials such as poultry excreta, poultry litter or vegetable matter, although containing some 70–80 % water, are essentially solid in form as is the traditional farmyard manure in which straw is mixed with cattle excreta. For digestion in the conventional continuous system these feedstocks must be mixed with water and in some cases chopped to make a pumpable slurry. It is possible to carry out batch digestions of the 'solid' materials, though, and such digestions were tested some years ago, but few data on performance are available. More recently, interest in the system has revived. The time course and the kinetic model of the 'dry' digester is essentially that of the batch culture shown in Fig. 3.1, and Jewell (1980) suggested that reaction times of 100–300 days would be required for total gasification of the biodegradable material.

THE ENGINEERING MODEL

In the previous section different types of digesters and some theoretical aspects of the systems have been discussed. From data based on small-scale experiments, with or without application of a biological model, and consideration of the type of feedstock and its availability, a continuous or batch-digester plant of suitable type may be selected. The next step will be, for instance, to determine the size of plant required if a fixed flow of feedstock (say excreta from a certain farm) is to be treated, and the results of this treatment in terms of utilisable gas energy, digested sludge solids or other parameter.

A determination of utilisable energy obviously requires a calculation of energy produced, and energy utilised in heating the digester at least. Energy produced can be calculated if the feedstock is defined and suitable experiments have been done. In some cases with farm wastes the actual flow of slurry and its composition in terms of total solids may have been determined and so from small-scale studies on farm wastes a fairly exact value for gas production from this slurry (and the detention time needed and so digester size) can be calculated. In other cases only the number of animals may be known and so some standard figure for excreta produced and its composition will have to be used. This may be quite different from the actual farm situation where washing water, rain water, or handling or animal housing methods, etc., may dilute or concentrate the slurry. The energy utilised, in a simple model produced just to get an overall idea of the results of the digestion, may be a general approximation. One often taken is that in a temperate climate about 30 % of the gas energy from a mesophilic single-stage digester is used in heating the digester and input. Using such an approximation the energy requirements for pumping input, mixing the digester contents and the pumping requirements of the heating circuit, are usually ignored as these are small compared with the figure for digester heating with its possible errors. However, this approximation supposes the feedstock to be at ambient temperature, a warm feedstock will obviously make a difference. Also, the figure ignores variations in the concentration of the feed; a slurry of 6 % TS will produce about twice as much gas per volume of feed and so of digester than a slurry of 3 % TS, but heat inputs will be the same for both slurries. A refinement of this factor would be to take actual mean temperatures of input during a year, or take summer and winter temperatures and from this and the volume of input to calculate the energy requirement for heating the input to digester temperature. The mesophilic digester in a temperate climate might then be assumed to have a fairly constant cooling rate; 0·4–0·9 °C per day is a typical figure used, and so from the volume the energy required to heat the digester can be calculated. In both cases the feeds can generally be assumed to be water for a continuous-flow digester. The metabolic heat from the digester bacteria is very small and can usually be ignored; it will only slightly decrease the cooling rate of the digester. A further refinement would be to assume a certain insulation of known heat transfer coefficient on the digester and, using this, to calculate the cooling rate of the digester at certain ambient temperatures. However, such calculations can be affected by climatic conditions of wind, rain, sunlight and so on.

The more refined these models are, the more exact will be the results, but even the simplest of models can be useful. A simple model will serve to show that the possible useful output of gas from, say, a small herd of cattle is nowhere near the energy requirements for heating the farmstead or running a milking machine, so that there is no need to go further in designing a digester for this farm as a producer of energy.

If the simple model shows that it could be worth further investigation, then the model may be further refined. The question will arise of what to do with the gas and how to heat the digester. The gas may be used in a boiler to provide hot-water heating for buildings or hot water for some process on a farm or factory. It may be possible to arrange a water circuit which will, under control from the digester temperature, divert water to heat the digester when necessary. Or a separate boiler system may be used for the digester heating-water. In these calculations boiler and heat exchange efficiencies will come in. An overall approximation of, say 70 %, for boiler efficiency may be used, or more exact calculations for a particular boiler set-up may be used, although this latter usually comes in during later design work. It may, on the other hand, be decided that the gas will run an engine to generate electricity or motive power for machinery. In this case engine heat can be used to heat the digester, and again approximations or more exact figures may be used. Some engines designed for running on gas and fitted with heat exchangers for cylinder block and exhaust suggest an overall figure of 75 % recovery of the gas energy; 25–30 % as electricity or motive power, 45 % as heat energy, but a new engine-generator model claims a recovery of 90 %.

On a very large scale, such as the digesters planned for utilisation of feedlot waste or vegetation of different kinds, it is proposed that the gas be purified and the methane fed into natural-gas distribution mains. Again models of such systems have used only the simple concepts suggested earlier or have used more detail in determining energy requirements for digester, moving feedstock, purifying the gas and so on. Such models may also need to bring in the energy requirements of planting, growing and harvesting the vegetation.

Some models have been used to try to show the possibilities of large-scale implementation of a new digester process or a standard digestion of a new feedstock, from the results of small laboratory-scale digestions. This type of model can suggest possibilities but may be rendered quite inexact by unknown difficulties in, for instance, large-scale handling of the feedstock or other engineering matters.

A further extension of the 'engineering' model is to determine not only uses of the gas but uses for the digested sludge. This model may require, for instance, assumptions or more exact detail of the energy costs for dewatering or drying sludge for fertiliser or feedstuff use, and if the latter, possibly an energy cost for bringing the material to suitable bacteriological standards for marketing.

Very many models of this type have been made and details published, and their value depends on the information put into them. Some ignore so many energy 'debit' factors in the process that they are wildly optimistic. A fault in some models purporting to calculate energy production on a 'country' scale from various wastes is to take figures for total waste production (often only calculated from average figures for some unit waste producer) ignoring the fact that many of the sources of waste are too small and too scattered to contribute to the overall production. Nevertheless, even models using limited information may indicate whether the proposed scheme has any merit at all. In general the model with limited input tends to give an optimistic idea of the possibilities of a digestion scheme and if it shows very little positive energy balance then it is almost certain that the scheme as proposed would not be viable. But if a scheme in a simple model shows reasonable promise of viability then, as implied before, this can be used as a basis for obtaining further information by small-scale experiments or for refining the model in the light of known facts.

By degrees, then, the initial simple model becomes a more exact model of the actual plant and will eventually become a full engineering design for a particular plant. In these latter cases full account of the uses of the gas will have to be taken into consideration and in a factory or farm plant this will involve investigations into the total energy and the types of energy used on the site and the best ways of replacing this either by the gas itself or by electricity generated from the gas. This in turn may feedback; to the digester heating, for instance. It may involve consideration of use of part of the gas as such and part for engine running, and the engine may be used for purposes other than electricity generation. For example, in some sewage works gas engines are directly coupled to compressors which supply air for sparged-aeration sewage water treatment.

However, at some stage or other the 'engineering model' develops into, or has added to it, the 'economic model', and this will be briefly discussed next.

THE ECONOMIC MODEL

The engineering model in all its stages is often used as the basis for an 'economic model'. This again has various complexities. The simple engineering model which gives the gas production from a certain amount of feedstock and the size of digester required for a certain flow of feedstock can, by making some assumptions about the cost of a digester system based on known systems, be used to calculate an approximate cost of gas production and suggest how economies of scale might influence costs, or what are the minimum digester and feedstock volumes to make gas production worthwhile when compared with the cost of natural gas, for instance. This type of model can, however, be in error in the costing of the digester plant. At the present moment there is little information on the costs of farm-animal-waste digesters or digesters built to process crop and other residues, because few have been built. The costs involved in building municipal sewage works digesters will overestimate the costs of farm plants. Only when the animal-waste plant reaches the proportions of those being built, or proposed, for feedlots of tens or hundreds of thousands of cattle will the construction methods and costs of the digesters involved (of one million gallons or more) become similar to those of municipal digesters.

Similarly figures for change in costs of digesters with increase in size taken from data on municipal digesters may be inadequate when applied to generally smaller and simpler systems.

However, if given reasonably correct input these models can begin to give some idea of the costs of digestions. Much more useful information can, of course, be obtained if the engineering model is more detailed and construction can be more adequately costed, and the more detailed the engineering information the more accurately can running costs be determined.

To the actual digester costs the costs of obtaining the feedstock may have to be added. This will be relatively easy to do if, for instance, the economics of collecting animal-waste slurries from a number of small farms in a neighbourhood are being considered, as the capital and running costs of slurry tankers used on farms are well known and these would be adequate for the system proposed. But it may be much more difficult to make reliable calculations if it is proposed to collect animal wastes over county or state areas to supply one big digester unit, as has been proposed at times. An area where detailed economic modelling is feasible is in the proposed systems for digesting sewage sludge and municipal garbage in existing sewage works or in extensions of these. Here the costs involved in the digester system and collection of the sewage and separation of the sludges are known as the systems exist. The costs of municipal garbage collection are also known, as are costs for sorting out metals, glass, plastics and other non-biodegradable rubbish. Costs of shredding garbage are also known and costs of slurrying the garbage with sewage sludge and pumping this, etc., can be adequately estimated. Small-scale tests have shown the best combinations of sludge and garbage and detention times to be used and gas productions to be expected, and from these plans for running of the full-scale digesters on a mixed feedstock can be made up.

These models in varying complexity can give estimates of costs of energy production. Even if not very accurate they can give an order of costs which can show if the costs are approaching those of conventional fuels and whether it is worthwhile even considering the process further. If the process seems possibly worthwhile then further improvement of the model, or further practical research work on a small scale, or the building of a larger-scale pilot plant to get more adequate information may be indicated. And it need hardly be said that the economic model also takes into account monetary inflation and trends in price rises or supply of other energy sources; what may be too costly today may be feasible tomorrow.

Eventually, of course, the economic model, like the engineering one, progresses to the cost estimates of constructing a plant to a particular engineering design and the costs of the energy produced by this actual plant. In the more complex model, costs of using the biogas or at least costs of purifying the gas if necessary (say for addition to existing household gas lines) will be taken into consideration. Here knowledge, in lesser or greater detail, of capital and running costs of internal combustion engines of various kinds, gas turbines, electricity generators, switch gear, and so on, will be required.

The economic model must finally consider the use or disposal of the non-gaseous products of digestion, and these may make a considerable contribution to overall cost balances. There may be a defined use for the products, say as fertiliser on an existing farm. In this case the distribution system will probably exist if untreated slurry has previously been used as fertiliser and it will be a case of showing that the digested slurry is easier handled than the original, and possible economies are involved in this. On the other hand if slurry is being collected from a number of farms or feedlots, or there is inadequate land available for nearby spreading, the cost of long-range sludge distribution or relative costs of carting whole sludge or dewatering it and distributing dried sludge solids as fertiliser may need to be analysed.

There may be no obvious outlet for a digested sludge so the economics of various alternative uses, say as animal feed, feed for fish ponds or algal growth, may be compared. Various factors may be involved here; for instance, if an animal feed is to be prepared for commercial sale it may have to reach various statutory bacteriological and chemical analysis requirements. Again, the economics of drying and separation of digester bacteria from undigested residues for sale of the former as a higher-protein feed than the unseparated sludge solids may be considered.

If sludge solids are separated, then the use or disposal of the water may need to be determined and costed. Disposal may involve further purification, in which case a further engineering model may be needed on which to base the economic model. And this brings in another factor in an economic model, control of pollution by digestion. The economic benefits of this may be difficult or almost impossible to cost adequately unless the producer of the feedstock is faced with a definite pollution problem involving possible closure of the farm or factory, or heavy penalties for causing a nuisance. However, increasing legislation in many countries about pollution control is making this factor of more defined importance as time passes.

As in the case of the engineering and other models, this has been only a brief summary of the structure and uses of the economic model. Very many of these models exist in published papers, and it would be difficult to select some just as examples; however, some examples illustrating various points will be found in later chapters of the book.

REFERENCES

CLAUSEN, E. C., SILTON, O. C. & GODDY, J. L. (1979). *Biotech. Bioeng.* **21,** 1209.

COHEN, A., ZOETEMEYER, R. J., VAN DEURSEN, A. & VAN ANDEL, J. G. (1979). *Water Res.* **13,** 571.

GHOSH, S. & KLASS, D. L. (1978). *Proc. Biochem.* **13,** 15.

HAMER, M. J. & BORCHARDT, J. A. (1969). *Proc. Am. Soc. Civ. Eng.* **95,** 907.

HENDERSON, C., HOBSON, P. N. & SUMMERS, R. (1969). In: *Continuous Cultivation of Microorganisms.* Academia, Prague. p. 189.

HERBERT, D. (1961). In: *Continuous Culture,* Monograph 12, Soc. Chem. Ind. London. p. 21.

HERBERT, D., ELSWORTH, R. & TELLING, R. C. (1956). *J. Gen. Microbiol.* **14,** 601.

HILL, D. J. & NORSTEDT, R. A. (1980). *Ag. Wastes.* **2,** 135.

HOBSON, P. N., BOUSFIELD, S. & SUMMERS, R. (1974). *Critical Reviews in Environmental Control* **4,** 131.

HOBSON, P. N. & MCDONALD, I. (1980). *J. Chem. Technol.* **30,** 405.

HOBSON, P. N. & SUMMERS, R. (1966). *Nature* **209,** 736.

JEWELL, W. J. (1980). In: *Anaerobic Digestion* (eds. D. A. Stafford, B. I. Wheatley and D. E. Hughes) Proc. 1st Int. Symp. An. Dig., Cardiff. Applied Science Publishers, London.

KIRSCH, E. J. & SYKES, R. M. (1971). In: *Progress in Industrial Microbiology,* Vol. 9 (ed. D. J. D. Hockenhull) Churchill Livingstone, London. p. 155.

LAWRENCE, A. W. (1971). *Adv. Chem. Ser.* **105,** 23.

LEVENSPIEL, O. (1972). *Chemical Reaction Engineering.* John Wiley, London and New York.

MILLS, P. J. (1977). In: *Proc. 9th An. Waste Manag. Conf.,* Cornell.

MONOD, J. (1942). *Recherches sur la croissance des cultures bacteriennes.* Herman et Cie, Paris.

PIRT, S. J. (1965). *Proc. R. Soc. Lond. Sect. B. Biol.* **163,** 224.

PIRT, S. J. (1974). *J. Appl. Chem. Biotechnol.* **24,** 415.

PIRT, S. J. & KUROWSKI, W. N. (1970). *J. Gen. Microbiol.* **63,** 357.

POWELL, E. O. (1965). *Lab. Pract.* **14,** 1145.

POWELL, E. O. & LOWE, J. R. (1964). In: *Continuous Cultivation of Microorganisms.* Academia, Prague. p. 45.

SHEIH, W. K. (1980). *Biotech. Bioeng.* **22,** 667.

TEMPEST, D. W., HERBERT, D. & PHIPPS, P. J. (1967). In: *Microbial Physiology and Continuous Culture.* Proc. 3rd Int. Symp., HMSO, London.

YOUNG, J. C. & MCCARTY, P. L. (1968). Tech. Rept. No. 87, Department Civ. Eng., Stamford Univ, USA.

YOUNG, J. C. & MCCARTY, P. L. (1969). *Water Pollution Control Federation* **41,** 160.

CHAPTER 5

Types of Digesters Being Constructed and the Operation of Digesters

Some theoretical aspects of digester design were considered in the last chapter and in this chapter some more practical aspects of digesters which have been built will be considered. Detailed engineering design considerations cannot be discussed in a book such as this, but some general methods of construction will be outlined.

Digesters vary in construction and design, but one thing is common to all, and that is that the digestion has to be started and the methods of starting are universally applicable, so this will be discussed first.

STARTING A DIGESTION

Although it is often said that start-up is the most difficult part of digester running and the time when failure is most likely to occur, experience with farm-waste digesters shows that providing some general rules are followed failure of digestion is very rare, and this also seems to be fairly general experience with domestic digesters.

As was outlined in a previous chapter, there are two methods of starting digesters treating faecal wastes—with and without an 'inoculum' or 'seed'—as the wastes already contain the necessary bacteria, although not in the correct numbers or proportions. With vegetable material or factory wastes an inoculum will always be necessary, as these feedstocks will contain few, if any, of the bacteria required in digestion. Most experience in digester running has been obtained with continuous-flow farm-waste digesters and municipal sewage digesters.

However, whether the digester is a continuous-flow or a batch digester, the digester tank can be filled only to its working height. With a floating-roof digester the roof can be lowered to give a minimum volume of gas space, but with a fixed-roof digester the gas space above the liquid level cannot be altered, although the gas-holder can be set to a minimum volume of dead-space by allowing a floating holder to settle or by collapsing a rubber bag or other mechanical holder. But whatever method is used there will still be air trapped above the digester liquid and in pipes and gas-holders and there is the possibility of forming explosive mixtures of gas and air as the digestion starts. To ensure absolute safety this air can be flushed out by purging the digester, pipes and gas-holder with an inert gas once the digester has been filled with liquid. On first starting their 13 m³ piggery-waste digester the authors' group purged the digester system with nitrogen (Robertson *et al.*, 1975). Purging with inert gas has also been advocated for municipal digesters (Safety in Sewers and Sewage Works, 1969). However, in later restarting of the piggery-waste digester or starting a companion digester or some later commercially-built digesters on farms it has been found that no purging is necessary. As previously explained, bacterial metabolism in animal wastes or sewage sludge will tend to use up oxygen in the digester head-space and so leave an atmosphere of nitrogen. With an inoculated digester, particularly, fermentation will begin quite quickly and the volume of gas produced will soon bring remaining oxygen in the head-space to a low percentage of the total gas. A suitable procedure with the farm-waste digesters is to allow gas production to say one third the volume of the gas-holder (either separate or floating digester-top), release this gas from the holder to atmosphere, and then to allow another two gas-holder volumes of gas to be formed and released before test-lighting the gas at a small pilot jet or lighting a small sample of gas taken into a football bladder or similar container which can be connected to a small jet. Since the relative volumes of digester head-space and gas-holder can vary with digester design some allowance should be made for this in determining the volumes of gas to be collected before testing the gas from a digester with separate gas-holder. However, if gas to about three times the dead-space volume of the digester system is collected and blown out of the system, then no dangerous concentration of oxygen should be left in the system. Similar procedures seem to be generally used in starting large municipal digesters (see for instance; A Symposium, 1964). If either laboratory or portable apparatus is available then gas analysis can be used as an additional safeguard before the gas is first lit. It need hardly be said that suitable precautions about lights and smoking should be observed when gas is being vented.

The actual start-up of the digestion can be accomplished in a number of ways. It is possible to start up a digestion from a digester filled with water. A small-scale piggery-waste digestion was started from a digester filled with water and heated to operating temperature (35 °C). Loading of the digester with a slurry of about 2·5 % TS, in volume equivalent to a detention time of 37·5 days, initiated good gassing in about 42 days and the digestion stabilised in about 63 days. The loading rate was then gradually increased in both solids content and volume of slurry to the operating parameters (Shaw, 1971; Hobson & Shaw, 1973).

This type of starting digestion from water where the digester is immediately started as a continuous culture should be distinguished from the recommendation of the Safety in Sewers and Sewage Works (1969) paper that a digester should be filled with water to reduce the air volume in the tank before the tank is filled with raw or seeded sludge by displacing the water with sludge pumped into the bottom of the tank. However, the recommendation does not appear to be generally followed and the procedure for starting a digestion from a filling of sludge or slurry can be generally described as follows. Only general indications of volume are given because actual practice has varied. Descriptions of some starting procedures for farm-animal-waste digesters and municipal digesters are given in various papers (e.g. Robertson *et al.*, 1975; Summers & Bousfield, 1976; Hobson *et al.*, 1979; A Symposium, 1964; Sambidge, 1972). The amount of seed sludge added depends to some extent on what is available. In the case of starting a municipal sewage digester amongst a group of digesters already running, the digester can be filled with digested sludge from the overflows of the other digesters. This may take a week or so, but it gives a 100% inoculum for starting. An inoculum from a working digester with the same feedstock obviously gives the best start to a digester, but this may not always be available. In general, digesters for the farm wastes and other feedstocks considered here will be single units and seed from a similar working digester will not be available. Farm-waste digesters may be started with a seed from nearby domestic digesters and if the farm digester is not large then an almost 100% volume of seed may be obtained using road tankers. Seeds for both farm and sewage digesters and for a digester treating palm oil waste (Morris, 1980) have been obtained from the appropriate waste which had been standing at ambient temperature in open tanks for some weeks or more and so begun to develop an anaerobic digestion. Such a seed may not need to be developed on purpose but may exist in the waste collection or treatment system used prior to the installation of the digester. The amount of seed used has varied; the authors have started pilot-plant farm-waste digesters with as little as 5% volume from a working digester (not necessarily the same feedstock, e.g. a cattle-waste digester started from a piggery-waste digester). Ten percent inoculum has been used with domestic digesters.

As might be expected, the greater the volume of active inoculum used, the quicker and more surely digestion will start. As previously mentioned, vegetable wastes and similar materials which contain few bacteria will need an inoculum, but digesters containing faecal wastes will start slowly without an inoculum. For starting a vegetable-waste digestion some farm slurry might be added if a seed from a working digester were not available.

The digester is filled with the feedstock, with or without the seed material, in dilute suspension; about 2–3 % TS has been used in starting farm-waste digestion, and domestic digesters usually start with about a 3 % TS sludge. When the digester has been filled it can be heated up to working temperature using some form of auxiliary boiler instead of the digester-gas boiler or engine which will finally supply heat. The heating process will take some days at least and may take a week or two, but in any case gradual heating up of the digester is best for the bacteria. Convection currents will give some mixing during the heating-up period, but even if not during the heating-up, at some time during the start-up period more vigorous mixing will be needed to prevent too much settling of heavier solids and floating of light ones. If a mechanical mixer is part of the digester design, or if heating is by external sludge circulation, then mixing will not be a problem, even if gas mixing is to be finally used with the external sludge heater. But if gas recirculation is the only mixing system fitted, then unless gas production starts in a few days, some form of temporary mixing with compressed gas or an external pump circulating the digester sludge will be needed.

Once operating temperature has been reached the digester should be left for a few days as a batch digester before loading at a low rate, say, equivalent to a 40- or 50-day detention time and with feedstock of low (2–3 %) TS. Gas evolution may start during the first few days if a good inoculum has been used. If gas evolution has not started then it may be better to leave the digester unloaded until some gas evolution is apparent. Once the digester is evolving gas then loading at the low rate should be continued for some weeks until the gas evolution has stabilised. The loading rate may then be gradually increased over a few more weeks to the operating conditions. It is difficult to give exact times for start-up, or for changing loading rates during start-up, of digesters. Small-scale farm-waste digesters have started gassing well in six days when inoculated from a working piggery-waste digester and the loading rate has been increased almost immediately. Other digestions have needed three or four weeks at low loading rate to give good gas production before being built up to working rate. Periods of two to four months have been quoted for the start-up period of municipal sewage digesters before they were put under working load. The main thing is to proceed slowly.

During start-up, gas production is a good guide to performance of the digestion. The gas volume is easily measured with the equipment on the digester, but gas composition may not be so easily determined, although its analysis is desirable. A simple burning test may be adequate as an indication of methane content (if it burns the gas is about 50 % or more methane) if gas analysis apparatus is not available. The carbon dioxide content of the gas may be measured by shaking a known volume of gas with a dilute sodium hydroxide solution and noting the decrease in volume. The residual gas may be methane or may be air or nitrogen. But production at increasing and then constant rate is a good indication that the digestion is proceeding satisfactorily. If the equipment is available the measurement of pH and total Volatile Fatty Acids (VFA) will give useful information. The VFA may increase initially but should soon begin to fall, and pH should move towards 7 or just above.

If the VFA increase suddenly, or the pH falls, or gas production falls, during the period of initial loading of the digester (and suddenly here means over a few days), or if these measurements show a general change in direction away from the trend towards low VFA, just above neutral pH and increasing or stable gas production, then loading should be stopped for a few days when VFA should decrease and pH return towards normal. Loading can then be restarted but at a rather lower rate until conditions seem stable again, when the rate can be increased. The addition of lime is advocated by some workers as a cure for a digester which is going acid. This has been and still is done, but it can lead to complications and should really only be done as a last resort if the pH persists in remaining low (say below 6·5) for some time or acids remain high and there is no sign of recovery even though loading has been stopped. It is difficult to give an exact figure for VFA levels which should be regarded as high. Experiments on the kinetics of methane production from acetic acid in laboratory culture and mesophilic piggery-waste digesters show that, as expected from the continuous culture theory discussed previously, the steady-state level of acid (the limiting substrate in this case) in a properly functioning digester varies with detention time (dilution rate). In a digester running at 20, 30 or more days, detention time levels should be about 100–200 mg/litre. At 10 days the level will be about 300–400 mg/litre and the level rapidly rises at detention times below 10 days (Hobson & McDonald, 1980). Experiments with pure cultures of a hydrogen-utilising methanogenic bacterium (Hobson & Shaw, 1976) showed that while acetic and butyric acids in any concentration were not inhibitory, propionic acid was partially inhibitory at 1000 mg/litre and inhibition increased with concentration of acid. While the residual acid in a properly functioning digester is entirely, or largely, acetic, partial failure of methanogenesis from hydrogen can tend to increase the proportion of propionic acid in the VFA. The acid further inhibits methanogenesis and the inhibition could become, in effect, 'autocatalytic' (Hobson *et al.*, 1974). The composition of the VFA in a digester may, then, be more important than the amount. However, propionic acid, as well as butyric and acetic acids, does seem to be converted to methane in digesters and cessation of loading may allow the situation of excess acid to correct itself.

Ammonium hydroxide has also been suggested as a neutralising agent for an acid digester, but apart from problems in handling, use of ammonium salts could lead to ammonia inhibition of digestion. Sodium hydroxide or carbonate or bicarbonate are expensive and difficult to handle. Ammonia inhibition of a starting digester has not been known to be a problem. Even wastes such as poultry excreta which contain large concentrations of ammonia do not give solutions of high enough concentration to be inhibitory when used in the low TS slurries recommended for start-up.

If a digestion does not respond to the stopping or slowing down of loading and appears to have completely ceased activity, and if this is known to be due to acidity, then the only alternative to use of lime (or if even this has failed to restore activity) is to partially empty the digester and add sufficient dilute slurry to reduce the acids to more normal concentrations. Additional seed slurry may also be needed. However, such extreme failure of digestion appears to be rare.

It has been assumed, of course, that there are no toxic materials in the feedstock. The same kind of feedstock will have been known from operation of other digesters, or from small-scale tests, to be digestible before the particular digester was built and a survey of the particular source of feedstock should have been done initially to ensure that it did not differ in any relevant way from that of feedstocks previously digested. For instance, while it is known that dairy-cattle wastes digest well, it could be that a particular milking parlour and dairy uses a different detergent for cleaning equipment than normal, or more detergent than usual gets into the cow slurry. The detergent may be toxic to the digester bacteria in these circumstances and so steps would have to be taken to eliminate it from the digester feed.

THE SINGLE-STAGE STIRRED-TANK DIGESTER

This is the most common type of digester and is the one almost invariably used for feedstocks of high (2–10%) TS content. The municipal sewage digester is of this type, as is the small, peasant-farm digester of the East. These two uses span the whole range of digester sizes from up to 4·5 × 10³ m³ or 1 million gallons (or in a few cases double this size) for the sewage digester, to about 2 m³ for the peasant-farm digester. The digesters being planned or built at present for energy-generation from agricultural wastes are generally from about 250 to 1500 m³ as this covers the usual range of intensive pig or cattle farms and the smaller poultry units. However, digesters of 4500 m³ have been built for very large cattle feedlot units in America and digesters of similar size would be needed for the bigger poultry and duck farms in Britain and elsewhere. If large-scale exploitation of energy crops, seaweeds, or other feedstocks were to come about, then digesters of this latter size would be the rule.

The size of a single-stage digester is, as explained before, a direct multiple of the daily volume of feedstock and the detention time. Although there are advantages of scale in the economics of tank building and in energy balances (a large digester has a lower surface to volume ratio than a small one, and so comparatively smaller heat losses), for large volumes of feedstock it is obvious that sheer mechanical problems may dictate that a number of digesters are required. Large municipal sewage works may have up to twenty or so million-gallon digesters. These digesters all take a portion of the feedstock flow in the same volumes and detention times and work independently, although all the gas is usually piped to the same engine-house. It has been suggested, however, that an advantage may be gained by having a number of smaller digesters linked in series rather than one big digester.

Theoretical aspects of the multi-tank digester have been previously considered, and while these may show some increase in production this has to be balanced, as previously mentioned, against economic and energetic disadvantages. A number of small digesters would be more costly to produce than one big digester. Heat losses would be greater from the greater surface area; each digester would need stirring and heating systems; pumps would be needed to transfer sludge from one tank to another. Unless digester production could be greatly increased, there would seem to be, overall, no advantage in the multi-tank digester. So far as the authors are aware a proper multi-stage digester system has not been tested in practice on a large scale.

The Small, Simple Digester

In developing countries there is need for small-scale, 'alternative' sources of energy. Wood in many areas is becoming scarcer, the use of dried animal excreta wastes a valuable fertiliser and increasing fuel prices are putting small oil and LPG stoves even further beyond the reach of the people. Biogas units would seem to be the ideal solution as they produce fuel from almost any organic material, and a fertiliser. Small biogas units can only be viable, though, when the energy inputs are minimal, so the countries where the small units are being built or considered must have a climate sufficiently warm to obviate artificial heating of the digesters, and the digester must be hand-loaded and -emptied. The first consideration does rule out a number of areas where although summer temperatures are high, temperatures in winter when most gas is required are too low for efficient digester running.

The greatest number of small digesters have been built in India, China and Taiwan. In India some thousands have been built and in China some millions, but what is not certain is how many of these are running, or running efficiently. The running of even a simple digester demands care and attention on the part of the owner and this can only be obtained by a programme of instruction in the running as well as in the building of digesters. Such programmes are being run in these countries. Research and development programmed towards improvement in the design and efficiency of the small digesters is also being carried out. Also being investigated are ways to make these digesters cheaper to build. Although their cost is minute in terms of the economics of a rich population, in terms of the economics of the people for whom they are designed the cost is often high, and in some cases cost is a factor limiting further spread of the systems. However, whatever the economic or other limitations, there is no doubt that small digesters are seen as a major source of energy for single houses or small communities in a large area of the world.

The Indian digester is generally based on the design instituted by J. J. Patel and is referred to as the 'Gobar' digester. The Gobar digester development scheme under the Khadi and Village Industries Commission has been instrumental in spreading digester technology, although others are also concerned. The Gobar digesters are built by many small firms and ancillary equipment such as gas rings, stoves and lights can be obtained. Besides popular literature on the digesters, a number of books and research papers have been published: for instance, Patel, 1951, 1959, 1963, 1964; Sathianathan, 1975. *Gobar gas—why and how*, 1975.

The Gobar Digester

The typical Gobar digester is a brick-lined cylinder sunk in the earth, with a wall dividing the cylinder into two, and with inlet and outlet ports leading to the bottom of the tank, as shown in Fig. 5.1. The top consists of a gas dome

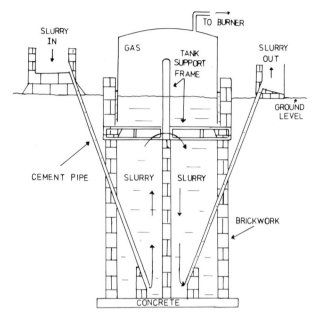

FIG. 5.1.   One type of Indian 'Gobar' digester.

floating in the digesting sludge or a separate water channel. Stirring is accomplished by convection currents and gas formation in the digesting sludge and by a paddle attached to the gas dome. As this latter rises and falls the paddle breaks up crusts and the dome can also be moved backwards and forwards by hand at intervals to break up surface scums. The gas dome is made of any convenient sheet iron. Gas is taken from the top of the dome via a flexible pipe to short runs of rigid piping leading directly to the gas burners of the house. Condensate pots in the gas lines are recommended.

The feedstock is usually cattle manure plus 'night soil', the human excrement. This runs, or is manually scraped, from the animal and human housing near to the digester. The output is taken off from the outlet tank and used as fertiliser either immediately or after some further settling or maturing. Thus, as is said in the literature, a fertiliser retaining the original plant nutrients of the excreta but with less weed seeds and pathogenic organisms, as well as the biogas, is obtained. Heating of the digester is purely by insolation and the generally warm air and ground. In some areas a fairly uniform temperature can be obtained, but in other areas, or at different times of the year, there are large temperature variations between day and night, and in some areas, as already pointed out, digester running

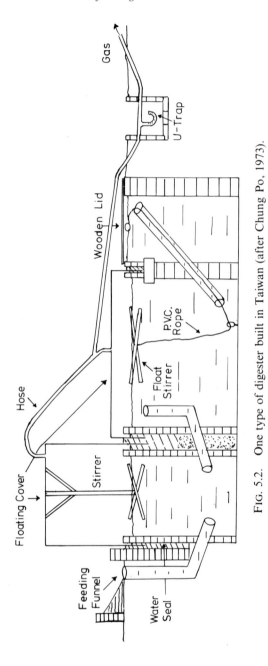

FIG. 5.2.　One type of digester built in Taiwan (after Chung Po, 1973).

in winter months is not possible. Digestion is slower than in the heated, stirred, high-rate digesters and detention times of about 50 days seem to be general. The gas produced contains about 55 % methane and some figures suggest that total gas production per unit weight of cattle manure added to the digesters can be equal to that quoted later for experiments in the West. However, there must be variations and, particularly as the diets of the cattle are often poor, gas production per animal must vary and could be considerably less than from the highly-fed animals referred to later.

The most common size of digester is that for the small, peasant farm and a volume of about 2 m³ is suggested for two to three cattle. This is the smallest possible size and bigger digesters are probably a more viable proposition. Bigger digesters are built for bigger farms and for institutional purposes where a volume of about 2 m³ for 60 people is suggested. The gas requirements for an Indian family have been given as 8 ft³ (0·22 m³) per person per day for cooking, and for lighting 4·5 ft³ per lamp of 100 candle power (*Gobar gas—why and how*, 1975).

Digesters of the sunken tank, floating gas dome type similar to the Gobar digester are being tried in many parts of the world. In Taiwan, digesters are similar in principle, but the recommended types (Chung Po, 1973) have either one or two rectangular brick chambers, the latter being inter-connected, with a floating gas dome on the first chamber (Fig. 5.2). The main source of animal excreta in Taiwan is pigs and for a digester as in Fig. 5.2 used on a 20 pig farm, the size of the first chamber is 3 ft 8 in × 3 ft 8 in × 6 ft deep and of the second 8 ft 2 in × 4 ft 10 in × 6 ft deep. The third chamber is only for storage of digested sludge and can be any convenient size. The digesters are manually operated, stirring being by a rope-pulled paddle. Emphasis is laid on use of the digested sludge in small, shallow ponds for growth of algae. Although the digesters are simple, it is probably true to say that on the whole they are designed for use by much richer farmers than those of India. The cost of the one or two-chambered digesters was put at 3500–5500 Taiwan dollars (NT $) in 1973.

The Chinese Digester

It is only comparatively recently that reports on the Chinese biogas programme have become generally available (see *A Chinese biogas manual*, 1979; Su-Te Shian *et al.*, 1979). The Chinese have, again, concentrated on small, cheap digesters using animal and vegetable wastes, and producing enough gas for use on a small peasant farm, though some bigger digesters (*c.* 60 m³) have been built to a similar design and digesters for communal farms and villages are a part of the overall plan.

The digesters vary somewhat in shape, but unlike the Indian digesters all are below ground and have a gas reservoir of fixed volume, hence the gas pressure varies. A typical cross-section is shown in Fig. 5.3. Construction varies with the type of ground. In soft rock the digester can be cut directly out of the rock. In some areas the digester pit may be dug into soft ground and then lined with stone; in other areas the lining is of brick. The top can be a brick dome, or cast cement. But whatever the method of construction

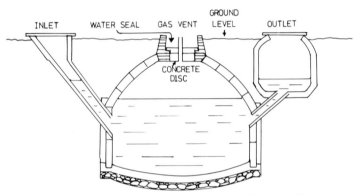

FIG. 5.3. Chinese type of digester. Construction varies according to ground conditions. The 'gas vent' is connected to the house gas-piping.

emphasis is placed on making the whole digester water- and gas-tight by application of layers of cement or mortars of various types. The top plug must also be well constructed and be gas-tight. Loading is by hand down one side pipe and unloading is again by bucket up the other pipe. The book referred to above, which is a translation of a Chinese instruction book on building and running digesters, says that loading must be regular, and the time-table for loading, once decided upon, must be strictly adhered to. One might expect that as the digesters are climatically heated, gas productions will vary with season and with location of the digester. As there is no stirring, some scum problems might be expected, and since instructions on the safety methods to be taken before entering a digester are given, some desludging at intervals as well as repairing of cracks are probably expected. A personal report to the authors from another area where building of Chinese-type digesters is being tried says that great problems with cracking due to small earth movements are being experienced. Meticulous attention to detail in building and running seems to be a main point of the Chinese digester programme, and the obtaining of advice from those who have built successful digesters is also stressed. This attitude could help to overcome problems.

The digester user is instructed to watch the U-tube manometer on the top of the digester and use gas only when the pressure is between certain marks. Instructions for construction of simple gas lights (mantles seem to be available) and gas-ring cookers are also given.

The examples shown in the diagrams are only representative. Many variations on these designs have been, or are being, constructed in the three countries named or in other warm countries with a small-holding farming system. In other areas, native workers and people from Western countries concerned with overseas development are also investigating the use of small digesters built from local materials or discarded materials (metal drums, etc.) from construction or other work. There is no doubt that digesters will work on a small scale and that they will provide gas for cooking and lighting for peasant-style living. However, even though these digesters seem simple and cheap, their cost is still in many cases high compared with the money available to the peasant farmer. Search for even cheaper construction is thus, in part, the reason for so many designs being considered. Butyl-rubber bag digesters were suggested as a more or less universal design a few years ago, but they do not seem to have been accepted. Butyl-rubber bags are not really cheap and there are problems in designing and running bag digesters.

Even the simplest of digesters demands some care in running, with attention to regular loading and clearing of blockages, etc. The single-farm digester may not always get this attention and so may fail. Bigger, village or communal farm, digesters may help to overcome these problems. A big digester should be cheaper to build than a number of small ones and its running can be better controlled by a properly instructed worker. Such units are now being considered or built in various areas.

Although the simple digesters described above can be increased in volume without change in design, at some point it becomes necessary, as well as advantageous from the point of view of increased efficiency of gas production, to add stirring systems. Although no stirrer is indicated in the reports, one might expect that the $60\,m^3$ Chinese digester previously mentioned would need some stirring system to prevent scum formation and sludge settling. In Japan, for instance, experiments are being conducted with tank digesters, situated in some cases below animal housing, which are stirred by one or more electrically-driven submersible pumps set on the bottom of the tank (Haga *et al.*, 1979). The gas production from the bigger digester can not only cover the energy cost of some form of stirring system, but also the energy cost of heating the digester. Even though climatic heating would still be the main input, the increased and more regular gas production resulting from proper temperature control could more than outweigh the use of some of the gas for heating.

As said at the beginning, there is no doubt that digestion can provide a convenient source of fuel in the less-developed countries, but there is still much work to be done in development of small digester designs and in instruction of the peasant farmer in the running of digesters before the full potential of the process can be attained.

The Large, Automated Digester

The Sewage Digester

The digesters used in sewage works are of two types. The oldest type, still used in some works, is an unstirred unheated tank, with (in cool or cold climates) detention times of some months. Layering of heavy sludge to the bottom, and lighter materials to the top of these digesters does not help efficiency and some zones may be more or less devoid of microbial activity. Gas production is slow and gas is generally not used. The purpose of the digesters is to produce a stabilised sludge and thus reduce pollution.

This type of digester has been superseded by the modern, 'high-rate' digester. 'Modern' is here a relative term as some high-rate digesters have been running for 40 years or so. During this time sewage digesters in general have had periods when they were regarded somewhat unfavourably by sewage-treatment engineers. During the 1950s introduction of some of the new detergents and trade wastes to the sewers caused toxicity problems to digesters as well as aerobic sewage-treatment plants. Production of biodegradable detergents and stricter control on discharge of trade wastes to sewers has largely overcome these problems. It must also be admitted that the running of digesters has not always been to a good standard, and some of the suggested instability and poor performance of digesters might be traced to irregular loading and so on. Properly run digesters in sewage works have been functioning without incident for years.

However, the realisation that digesters can be successfully run, together with the fact that alternative methods for getting rid of primary and excess aerobic sludges such as incineration, wet-air oxidation, centrifugal- or filter-dewatering, have had problems and have proved very costly, have brought digesters more into favour. And an additional point in favour of digesters over the last few years is that while the alternative methods mentioned require large amounts of energy for running, properly-run digesters can provide not only energy to run themselves but energy to offset a portion, or in some cases virtually all, of the energy required by the whole sewage works. Another factor in favour of digested sludges as against dewatered, but otherwise untreated, sewage sludges is the better properties of the digested sludges for land-spreading as fertiliser. Indeed, some reports (e.g. ESCA, 1976) recommend that because it is better with respect to pollution and content of pathogenic organisms than untreated sludges, only digested sludge should be land-spread. So apart from those already in existence, digesters are being built in new sewage works, and as populations increase more digesters are being added as older sewage works expand.

The high-rate sewage digester is heated, usually to about 30 °C, although some digesters run at lower temperatures, and is stirred to obtain even suspension of the sludge particles and to prevent scum formation. Thermophilic sewage digesters are very uncommon, for reasons which will be found later in this chapter when the section on building digesters for energy production will also include some discussion of the principles behind digester construction and running.

The size varies from about 120 000 UK gallons (540 m^3) for a small town of five or six thousand inhabitants to about a million gallons (4500 m^3), of which size there can be twenty or so, in the large-town sewage works. The digester size is based on detention times usually of 20–30 days, but in some modern plants detention time has been brought down to about 12 days. A sphere is the shape having least surface area to volume ratio and so is most favourable from the point of view of heat losses to the surroundings, but a large spherical tank is difficult and costly to construct. An ovoid shape can have some advantages and there are some digesters of approximately this shape. Some are in use in Germany and a detailed discussion of digesters constructed on similar lines in Los Angeles (USA) is given by Nelson & Bailey (1979). However, in general, engineering problems and cost dictate that the majority of digesters are cylindrical in shape. A tall cylinder has some advantages from the point of view of mixing the contents, and small digesters often have a height to diameter ratio of 1·5:1 or greater. Engineering problems in building large digesters of this shape, together with the question of reducing surface area, however, result in most large digesters being built as squat cylinders with a height to diameter ratio of near 1:1, or less than 1:1.

The usual material of construction for domestic sewage digester sides and bottom is concrete. This has both good constructional properties and also, in the thickness used in digester tanks (6 in to 1 ft) gives reasonable insulation against heat loss. By suitable surface treatment it can also be made waterproof. While digester tanks are built entirely above ground it is quite a common practice to sink the tank about three-quarters or rather more of its depth into the ground. This is done from the points of view of both additional insulation and lessening the visual intrusion of large tanks on sites which are often very near housing areas. Where the ground conditions do not permit excavation the tanks often have earth built up to the sides, and these earth banks are then grassed over. However, if digesters are built partly underground then access must be provided not only for pipe runs but also valve chambers and other control and inspection points underground alongside and below the digester tanks.

The bottom of the digester tank is of similar concrete construction to the sides. The bottom can be flat, but is often conical in form to allow grit from the sludge to settle to a central well from where it can be removed through a large-bore valve and piping. The slope of the bottom varies considerably; Brade & Noone (1979) in a survey of digesters in Britain found bottoms with as much as a 62° downward slope. However, in general, the conical bottom is much less acute than this. Occasionally, digesters have been built with a mechanical scraper to move grit towards the centre of the conical bottom, but generally movement of grit is left to gravity.

The construction of the main body of the digester in concrete is a fairly standard procedure and follows general engineering practice in reinforced-concrete tank construction. But although principles of digester design and running equipment are known to all and are very similar in all types and sizes of digesters the detailed application of these principles may involve designs peculiar to one firm of digester builders. So in the following descriptions and diagrams (Fig. 5.4), only general outlines of equipment will be given, with the occasional reference to a particular firm's products given as an example. Such references do not mean that the firm is the only one producing equipment on this principle, or that the equipment is the best. The authors are, also, most familiar with British practice, and so any reference will generally be to British equipment. Indeed, it is very difficult in digester technology to point to one piece of apparatus and say that this is the best, or only, solution to the problem. The feedstocks and digested sludges are, particularly now when so many feedstocks for digestion are being considered, so diverse in physical properties that trial and modification is the only way of selecting the final equipment to be installed.

However, even if tank construction is fairly uniform, differences in principle and design begin to appear in the digester top. This is generally of sheet steel or reinforced concrete and can conform to three basic designs, although all tops are of flat dome or conical shape for reasons of strength.

The fixed-top digester has a top plate integral with the digester sides. This is simplest in construction, but because the head-space above the digesting

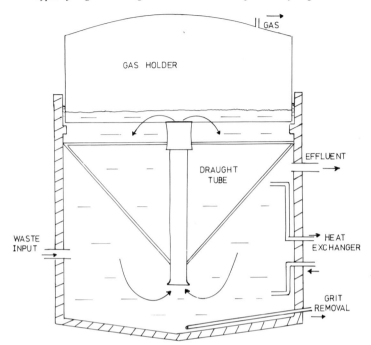

FIG. 5.4. Generalised municipal sewage digester with floating-top gas-holder. The draught tube can be mechanically or gas-operated and there may be more than one. Hot water heating pipes (heat-exchanger) may run to the draught tube or to separate coil or plate heat-exchangers in the digester. The effluent output pipe is connected to some form of weir or sludge seal to compensate for gas pressure in the holder, or in large digesters a pumped effluent discharge system may be used.

liquid is of fixed volume the digester must have a separate gas-holder to regulate gas pressure.

The other most common design has a top built with a skirt which acts as a gas-holder in the same way as the gas domes of the Indian and similar digesters mentioned previously. The skirt projects either below the digesting sludge or into a separate annular water trough to form a gas seal. As the gas is produced and utilised the 'floating top' rises and falls to give a varying volume of gas at constant pressure. The pressure of gas in the holder is usually about 6 in to 10 in WG, which is about that of the old town-gas supplies. Such a pressure allows for piping the gas to site of use and the use of the gas in burners or engines without complicated pressure-reducing valves. Even if the digester top is floating in the digesting sludge little gas is lost from the annular space as the sludge here tends to crust over or become covered with rain water and there is only a small amount of microbial activity. The floating top itself should be suitably stabilised to prevent its tilting as it moves, and this can be done either by a framework (on which run rollers attached to the top) built above the top edge of the digester-tank sides, or the floating top can move inside the digester sides which are suitably extended above the level of the sludge to form, again, a framework on which guide rollers can run.

Since the digester gas is used continuously in a sewage works and also because of the obvious practical difficulties of building a large floating top on a digester, the gas volume enclosed by the floating top is only a small fraction of the digester volume and can contain only a few hours' gas production.

If gas production is greater than gas usage and the floating top reaches the highest point of its travel, then the system acts as a safety valve and gas can blow out under the skirt and through the annular sludge. To confine this gas blow to one place and prevent disturbance of the top, some digesters have a 'notch' cut in the bottom of the skirt to allow a blow-off at slightly less than the general pressure imposed by the depth of submersion of the skirt.

If all the gas is being used then the top is prevented from dropping too far by stops, but since the gas is not generally being sucked out but being forced out by the pressure in the holder it is not likely that air could be sucked into the holder through the sludge around the skirt. However, if discharge and charge of the sludge do not keep in step then it is possible for the level of sludge to be brought below the level of the skirt. Such a happening is unlikely, but another form of floating top is designed not to cope with varying gas volumes, but with varying sludge volumes. This is exemplified in the Ames–Crosta Simplex system. In this case the top is combined with a skirt sealed in the digesting sludge, but the top moves up and down on guides which extend over a large proportion of the depth of the tank. The effective sludge volume of the tank can thus be varied over a large proportion of the total tank volume while the head-space in the digester remains the same. A separate, small, water-sealed gas-holder is then used to regulate the gas pressure. This system allows the digester to be used as a sludge 'holder' if required, i.e. instead of a constant volume of sludge being removed each day to secondary tanks where it can be stored until disposed of, the digested sludge can be accumulated in the digester for some time and then a large volume removed when disposal is convenient. Run in this way the digester would approximate more to a 'fed-batch' culture than a continuous-culture system. The digester efficiency will vary with the time for which sludge is held compared with the normal detention time. The theory of the 'fed-batch' culture differs from that of the true continuous culture and a theoretical model has been mentioned previously.

The digester top usually accommodates an inspection port with a heavy glass window and an internal 'windscreen-wiper' moved by a handle passing through bearings in the glass, and used to clear sludge spray from the glass.

Sewage digesters, with the exception mentioned above, are run theoretically as continuous-flow systems, but how continuous is the flow, varies. The input is pumped to the digesters from the primary sludge settling tanks and from the settling tanks for excess aerobic sludge. Since detention times in digesters are usually long, only a small fraction of the digester volume is added each day. As large-bore pumps are needed to pass the sludge, pump-flow capacity will generally exceed that required for the completely continuous loading which is theoretically the optimum, and pump-fed digesters are almost always loaded intermittently. The time between loadings may be only an hour or two, but in many cases it has been the practice to load digesters only every few days with a relatively large volume of fresh sludge. Such very intermittent loading may have contributed to poor performance of some digester systems. A recent survey of some digesters in Britain (Brade & Noone, 1979) showed that one digester was continuously loaded, but that loading of the rest varied from one to five feeds per day. Digesters in Los Angeles were loaded 17 times a day. The sewage sludge was continuously pumped to the 12 digesters and the flow was switched from one digester to another. The duration of the feed cycle also varies from some minutes to a number of hours. Centrifugal, scroll and stator, and piston type pumps are used for loading of digesters. The type used depends to some extent on volume being pumped, for instance centrifugal pumps have generally too high a flow rate for use in the smaller digester installations if anything approaching a continuous-flow feed system is to be used. The input pipe is below the level of the digesting sludge, so that the incoming sludge mixes with the active material. In some cases the input pipe discharges into a sludge-circulation mixer return pipe.

The overflow from digesters is generally by gravity discharge of some kind. The gravity discharge must, of course, incorporate a weir or similar system to compensate for the gas pressure in the digester by allowing a suitable head of sludge to build up and remain while overflow takes place. The use of a weir overflow provides a constant level of sludge in the digester since if the weir is of correct size the input is compensated by outflow over the weir and the outflow cannot reduce the sludge level below the lip of the weir. The height of the weir or the stand-pipe system must, of course, be commensurate with the gas pressure produced by the floating top or other gas-holder. Although gravity discharge is generally used, some digesters are operated on a pumped-output basis. This is often the case when loading takes place only every few days, and a digester used with various sludge volumes as previously described must have a pump-discharge system. Of course, a digester may be made, as for instance the experimental digesters of the authors and colleagues are, for long-term operation at different sludge volumes and here a number of weir-type overflows can be built into the digester at different heights, the weirs not in use being blanked off. Or a stand-pipe overflow of adjustable height may be used.

The 'high-rate' digester is stirred, as previously mentioned. The object of stirring a digester is to ensure that the digester contents remain relatively homogeneous and input sludge is mixed into the digesting sludge and also to ensure that the contents have an even temperature. The stirring is not required, as in an aerobic microbial culture, to mix a nutrient (oxygen) into the culture medium. Stirring of a digester, then, does not require the vigorous action or the very high power inputs of stirring an aerobic industrial culture. Stirring (or 'mixing' as it is generally and perhaps more properly called) of domestic sewage digesters is generally intermittent and is usually designed to turn over the digester contents every few (two to four) hours. Since the digester contains solid particles of various sizes and densities, and also in domestic sludge, fats and oils, there is a tendency for lighter particles and liquids to float to the top and form a surface scum and for heavier particles to sink. As mentioned before, grit can be allowed to collect to some extent on the bottom of the tank where it can be drawn off via grit-valves, but if there is no degritting system or grit accumulates too fast then the only solution is to close down the digester, empty off the liquid and remove the grit. It is obviously undesirable for this to be necessary more than once every few years (some digesters are closed-down for overhaul every five years), and grit accumulations can interfere with digester working, so the stirring system should circulate small particles of grit to be discharged with the liquid overflow. However, of more immediate effect than the slow accumulation of grit can be the comparatively short-time formation of a surface scum which can seriously interfere with the working of gravity overflows and prevent escape of gas from the liquid. So to prevent flotation and settlement as far as possible, mixer systems are generally based on provision of an up and down circulation of the sludge combined with a surface turbulence or discharge of sludge over the liquid surface which wets and sinks solid particles floating to the top.

Basically there are three methods of mixing, but there are many mechanical set-ups used. And since heating and mixing are often combined, the two subjects are best considered together.

As previously mentioned most mesophilic digesters operate at about 30 °C, but whatever the temperature, stability of this factor is important, and poor performance of some digesters may be attributed to poor temperature control. Heating is generally intermittent, being thermostatically controlled. Although some sewage digesters in the survey conducted by Brade & Noone (1979) were heated for 24 hours per day, some were heated for only 4 hours per day at the same season of the year. However, extent of heating time required will vary with the efficiency of the heat-exchanger system, as well as, obviously, with local climatic conditions.

Heating is most usually by hot water, used either directly or indirectly, although steam injection has been suggested, if not used (e.g. Brade & Noone, 1979). Steam injection heating of 9650 m³ sewage digesters in Los Angeles was reported by Garber *et al.* (1975). They said that the saturated steam, injected about 1·5 m below the sludge surface added about 3·4% more water to the sludge each day. The hot water is provided either by boilers heated by burning digester gas or by the heat recovered by heat-exchangers fitted to the cylinders and exhausts of digester gas or dual-fuel engines, or turbines, in the works' power station.

Indirect hot-water heating is by means of sludge recirculation. Sludge is pumped from the digester either intermittently or continuously through a heat-exchanger. The heat-exchanger systems used are of various designs. In some the hot water from the boiler or engines circulates through a shell-and-tube heat-exchanger or a coil-in-tank exchanger. The flow of sludge through the heat-exchangers is rapid and good heat exchange can be obtained with no possibility of the sludge caking in the pipes. The rapid flow of sludge can also either partly or completely mix the digester contents as inlet and outlet pipes are usually taken from different positions in the digester. However, many external heat-exchanger systems also need some other form of digester mixing. One external heat-exchanger system which is specifically designed for mixing is the Dorr–Oliver type B where the heat-exchanger is fitted on the outer wall of the digester. Sludge is withdrawn from part-way up the digester wall and also, as required, by an inside tube and funnel from the top of the liquid where any scum will tend to form, and returned to the bottom of the digester via a vertical heat-exchanger. Other types of external sludge-circulation heat-exchangers fitted to the digester wall are also manufactured.

An external heat-exchanger which uses, in effect, an integral boiler, is the type where the sludge circulates in pipes through a water bath heated by horizontal fire-tubes with digester-gas burners in the end. This type is used with the Ames–Crosta Simplex digester previously mentioned. But most heat-exchangers derive hot water from a separate boiler of the types used for natural gas but with slightly modified burners. In case of failure of the digester gas production, or for starting up or for periods of exceptionally cold weather when digester gas production may not be sufficient for all heating requirements, the boilers themselves are sometimes capable of being used with liquefied petroleum gases or natural gas with different burners, or auxiliary boilers fired by these fuels or oil can be brought into action. Works using waste heat from the power house generally have either auxiliary digester-gas boilers, or other-fuel boilers to compensate for an engine being out of use, or the type of problem referred to above. Gas boilers are about 60–80 % efficient in transferring energy from fuel to water. About 30 % of the heat energy of the fuel can be removed from a gas-engine cylinder-block heat-exchanger, and with an exhaust gas heat-exchanger this recovery is normally raised to about 45 %. With about 28 % energy recovery as electricity generated by an engine alternator system, such heat-exchange gives a total energy recovery of about 75 %. However, a recently introduced small (15 kW) gas engine unit, with engine generator and heat-exchangers totally enclosed, claims an energy recovery as heat and electricity of 90–92 % of the fuel gas energy.

The hot water from boiler or engine is also used in heat-exchangers situated in the digester contents. Such a system replaces the large, energy-consuming sludge recirculation pumps by the small pumps required only to circulate hot water through the digester. Some earlier systems used hot-water pipes situated around the wall of the digester, but these pipes were very liable to become coated with sludge, which eventually caked on to the hot pipes and so their heat-exchange efficiency was much reduced. Prevention of sludging-up of heat-exchangers is a vital part of in-tank heater design and one of the reasons why heat-exchange and mixing are often combined.

The objectives of mixing of digester contents have previously been described. While it is possible to mix a small digester by means of one or more paddle or flat-blade turbine stirrers, it is almost impossible to mix large volumes of thick sludges by this method. However, mechanical stirring is used in large sewage digesters. This takes the form of a screw pump which moves the sludge up a draught tube (Fig. 5.4) and throws it out either at the surface, or just below the surface, of the digester contents. In some designs the rotation of the pump can be reversed to suck a surface scum down the tube and out into the bottom of the tank. In this case the pump may be programmed to work alternately up and down or may be set to work downwards at longer intervals, once a day perhaps. Depending on the size of digester there may be one or a number (perhaps four) of these draught-tube mixers fitted.

Mechanical mixing by means of sludge circulation through an external heat-exchanger has already been mentioned, and sludge-circulation mixing can also be used without a heat-exchanger. In some cases the sludge input pipe is so arranged that the feed sludge mixes with circulating sludge.

Gas mixing is now becoming very common in sewage digester plants. In these systems digester gas is taken from the digester head-space, compressed to the pressure required to overcome the hydrostatic head of digester contents, and released at the bottom of the tank. In some systems the gas is allowed to rise freely from nozzles at the bottom of the tank, this causes turbulence at the top of the sludge and breaks up any surface scum, but is not so effective in lifting particles from the bottom of the tank. In another free-rise gas stirring system the gas is injected through the raw sludge inlet pipe, and in some cases the sludge withdrawal pipe has been used.

To get increased lift and circulation of the heavier sludge particles the gas can be channelled through a draught tube of design similar to that previously mentioned for the mechanical draught-tube mixer. In some designs the tube, gas jet and supply of gas are arranged so that during operation a continuous stream of small bubbles rises up the tube carrying sludge with it. In other cases a large bubble is formed at intervals in the tube and in rising carries up the sludge. Verhoff *et al.* (1974) have discussed the theory and optimisation of such gas-lift systems.

One or a number of concentrically-arranged draught tubes may be used in the tank depending on the size of the digester. However, gas-lift circulating pumps on the draught-tube principle and built externally to the digester tank are used. These, as in mechanical external circulation, take sludge from the bottom of the tank and throw it out near the top of the tank.

As with mechanical stirring systems, gas stirring is usually intermittent. It is impossible to give figures for either period of mixing or gas volumes required. Figures quoted from working digesters show total daily mixing times of some 4–18 hours, but in many cases it is not known if mixing was adequate. Even if mixing is to give homogeneous tank contents the time of mixing will still vary with the characteristics of the sludge being digested and the type of gas mixers (free-rising or draught-tube) used and their number and positioning in the digester. Gas volumes used also vary considerably, for the same reasons as times of mixing. Some volumes quoted by Brade &

Noone (1979) were 0·016 to 0·15 m^3/m^3 digester volume/h in the USA (Wiedemann, 1977) and 0·034 m^3/m^3 volume/h in Britain (Finch, 1956). Gas flows of about 0·15 to 0·025 m^3/min/m^2 of digester floor area are suggested by some British makers of gas pumps for intermittent gas stirring. Mills (1979) said that domestic digesters used gas flow rates in the range of 0·02–0·03 m^3/m^2 digester area/min. Garber *et al.* (1975) reporting on experiments on thermophilic and mesophilic digestion at a Los Angeles sewage plant gave the gas volumes for mixing digesters of 9650 m^3 capacity. The gas volumes used were 8·4 to 16·8 m^3/min injected through a number of discharge nozzles located near the centre of the conical bottoms of the digesters. The gas flow was free rising without draught tubes. This gas flow is about 0·001–0·002 m^3/min/m^3 of digester volume. Presumably this was a continuous gas flow.

There thus seems to be a wide divergence in gas volumes used for mixing digesters. This could be due to a number of reasons. It is not certain if the digesters quoted were being optimally mixed, under-mixed or over-mixed. The pattern of gas distribution would vary between digesters and some would have draught tubes while some would use free-rising gas. And the gas flow would vary from continuous to a few minutes per hour. Also the sludges would vary in physical properties. It is also difficult to give figures for the power used in gas or mechanical mixing as reports vary considerably. The power input to compressors or stirrer motors will vary with the type of unit used, and overall, of course, the energy input will vary with the time-on of the mixer system.

There are a number of papers on theoretical aspects of mixing in anaerobic digesters, the paper by Verhoff *et al.* (1974) previously mentioned, is an example. Different types of compressors are used for gas recirculation mixing. The digester gas contains some hydrogen sulphide and ammonia and these can have a corrosive effect on compressor components. The liquid-ring type of rotary compressor helps to combat corrosion by water-washing the gas as it is being compressed, and it also cools the gas and condenses water vapour out of it.

Some mixing of digester contents is also caused by the natural gas evolution and a well-fermenting sludge will help the stirring provided by mechanical methods.

Free-rising gas mixers can be combined with coil or other forms of hot-water heat-exchanger in the digester tank, the heat-exchangers being so arranged that the rising gas prevents sludge caking on the heater surfaces, but draught-tube mixers, whether mechanical or gas, can also function as heaters. The draught tube is built with a double skin through which hot water circulates. The Simon–Hartley 'Heatamix' is an example of a gas draught-tube mixer fitted with a hot-water jacket. In some cases a design with multiple draught tubes in a single water jacket has been used (see later). The heat-exchange capacity of such a system is generally calculated on the area of the inner tube as this is kept clean by passage of the sludge. The outer skin of the tube may cake up with sludge solids. Heat transfer coefficients for about 30 to 1700 $W/m^2/°C$ are quoted for heat-exchangers, but how many sewage digester heaters are working efficiently in practice it is impossible to say.

There is no doubt that proper mixing to give a homogeneous sludge of uniform temperature and efficient dispersal of raw-sludge feed (i.e. to make the digester as near as possible to the theoretical continuous culture described previously) is necessary for optimum digester performance, Swanwick *et al.* (1969) found in a survey of British sewage works that in the majority of cases of inefficient digester performance, inadequate mixing was the most likely cause.

This has been a very brief survey of the design of domestic sewage digesters. It has been brief because to cover the subject in detail would require much more than the space available. Sewage-sludge digestion has to some extent developed by trial and error, and full-size digesters have been running over the same time period as small-scale experiments to show the optimum conditions for digestion and the effects of inhibitory compounds and so on. And while small-scale experiments may show the optimum conditions in terms of temperature, loading rates, etc., of digestion, they do not, as will be discussed later in the case of other feedstocks, show up the problems, or the solutions to these problems, involved in the large-scale handling of sludges. Because of the nature of digester feedstocks and the digesting sludge it is difficult to design heating and mixing systems with complete accuracy. Trial and redesign on the basis of performance has to be resorted to, and because the sludges vary it is difficult or impossible to get one design or running system that will be equally efficient in every digester. A modern digester, built as a complete unit, should obviously be the most efficient, but even here some testing of operating parameters is required to obtain best results. But in many cases sewage digesters are old and to try to increase efficiency they have been altered as new equipment or new principles for feeding, heating or mixing have appeared, or perhaps, for instance, extra mixers have been put in to help inadequate old ones. This type of development accounts to some extent for the varying figures for power requirements for digester running and for the many figures of gas volumes used for mixing, time of running of heaters and mixers, and so on.

Digester manufacturers have produced different designs of equipment and some equipment has been altered or modified by the engineers in sewage works in the light of their experience. Some digester equipment has been built at the sewage works or to the designs of the works engineers. However, as digesters continue to operate and new ones are built and tested it is probable that designs will tend to standardise more.

There is no doubt that performance of sewage digesters, considered nationally or internationally, is very variable. Even in one large sewage works, digesters may not all give the same results, but between works one will seem to have consistently good digestion while another will have poor digestion. In some cases this is not due to poorer equipment in one works but poorer and less scientific operation. While inadequate operation was once tolerated or, perhaps, no means of improving it was known to the operators, the present increase in scientific study of digesters of all kinds and the need for increased efficiency in use and production of energy together with public health requirements for decreasing pollution caused by sewage sludges, should lead in the next few years to a general improvement in the performance of sewage digesters.

THE CONTACT DIGESTER

So far, few large-scale contact digesters have been built, so few details of construction can be described. The digester itself is basically the same as any other stirred-tank digester and can be constructed, heated and mixed in the same way as the high-rate sewage digesters or the farm-waste digesters described in other sections of this book. The ancillary equipment differs, though, in that there is some system of separating the bacteria from the outflow and returning it to the digester tank. If the feedstock contains particulate material then some of this will be returned with the bacteria (Fig. 5.5).

Fullen, Steffen and coworkers in the USA were some of the pioneers of contact digester design and usage for meat-packing waste-waters, and after experiments with a laboratory-scale and then a pilot-plant treating 7000 US gallons (26·5 m³) per day a full-scale plant was built at an American meat-packing plant (Schroepfer *et al.*, 1955; Fullen, 1953; Steffen, 1953).

The complete plant contained two digesters, each treating a flow of 530 US gallons (2 m³) per minute on average with a maximum of 620 gallons per minute. The plant was commissioned in two stages, one digester being run and tested in the first stage. The detention time of the

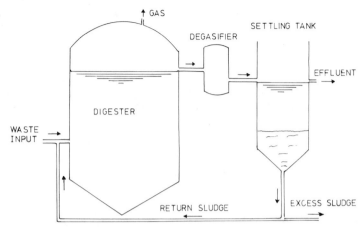

FIG. 5.5.   Schematic 'contact' (feedback) digester. The digester tank is fitted with
stirrers and heaters as in Fig. 5.4 (cf. Fig. 4.5).

liquid based on waste-water input volumes was 12·23 hours, but a return
flow of 2·4 volumes of settled sludge to 1 of raw waste decreased this
detention time to 3·58 hours.

Full details of the digester design are not given in the paper by Steffen, but
stirring was by 10 ft diameter flat-blade turbine impeller. The rotor plate
had 20 vertical fins standing 2 ft above and 4 ft 2 in below the rotor disc. The
disc was set 6 ft below the liquid level in the digester and rotated at 6·12 rpm.
The liquid depth in the digester was 21 ft.

As previously mentioned the difficulty with contact digesters is that
gassing in the digester outflow prevents particles from settling. The problem
was overcome by using a vacuum degasifier. Outflowing sludge entered a
distributor trough at the top of a steel tank 11 ft in diameter and 9 ft deep set
on the concrete digester cover. The sludge was drawn up into the trough by
a vacuum (20 in mercury) which also degassed the sludge as it trickled down
over slats in the tower. Gas coming off the degasifier was higher in carbon
dioxide content than the digester gas because of the release of carbon
dioxide from solution. The degassed sludge then ran to gravity settling tanks
from where the settled solids were returned to the digester via a suction
scraper rotating over the tank floor.

Hemens & Shurben (1959) described a different degasifier and sludge
separator built with a pilot-plant digester for meat-factory wastes in
Britain. This was a circular tank with a conical base, 6 ft in diameter and
1 ft 2 in deep at the side. A scraper arm rotated from a central shaft and,

fitted with scrapers set at an angle, moved the liquid around. Above the scrapers was a sheet-metal spiral baffle. Digester effluent entered at the centre and, constrained to a spiral path by the baffle, left at the edge of the tank. The scrapers moved settled solids to the centre of the conical bottom where they drained to a pump which returned them to the digester. The top of the separator was open and so gas removed from the liquid was lost to atmosphere. However, a full-scale plant would need some form of building around it as the workers found that separation was best when the digester effluent at 33 °C did not fall to below 20 °C in the separator.

In more recent developments, commercial plant treating various factory wastes uses a solids separator which cools the digester effluent to prevent gas bubbles forming.

An interesting use of a model (i.e. small-scale) in solving a mixing problem in one of these digesters treating wheat-gluten factory wastes is described by Irving *et al.* (1978). The digester tank (19 m diameter by 5·35 m deep) was stirred (heated) by six draught tubes arranged around, and close to, the centre axis of the tank with a deflector plate above. A total gas volume of 4·5 m^3/min passed through the tubes. The solids concentration in the digester tanks (i.e. in this case the bacterial mass) should have been 0·8 %, but soon dropped to 0·2 %. A 1:25 scale model of the tank and draught tubes was built and it was found that the flow pattern of the liquid was such that solids were being deposited at the outer bottom corner of the tank and on the bottom nearer the centre. The problem was overcome by enclosing the six draught tubes in a larger circular tube (3 m deep in the full-size tank) and putting six auxiliary spargers inside the big tube (in addition to the six draught tubes). This changed the flow pattern in the model and in the full-scale plant so that solids were not deposited and the desired uniform concentration of 0·8 % was maintained.

ANAEROBIC FILTERS

A full-scale anaerobic filter treating waste-waters from a starch-producing plant was operating successfully in America and was described a few years ago (Taylor, 1972). In this plant the waste-water was heated to 32 °C by steam injection before passing to the upflow filter. The water was distributed by a sparger in the tank bottom. Three filters were in use each treating a waste flow of 90 US gallons (0·34 m^3) per minute. The filters were wooden-stave tanks 20 ft high × 30 ft in diameter filled with broken rock of 2–3 in at the bottom graded to 1–2 in in the top half. The total volume was

300 000 US gallons (1140 m^3) and the void volume was estimated as 120 000 gallons (454 m^3). The filter effluent was passed directly to the sewers, although tests showed that a 30 min settling could improve the effluent, reducing the Suspended Solids from 1300 mg/litre to 370 mg/litre.

A pilot-plant contact digester (or sludge-bed digester) combined with an anaerobic filter was successfully tested for sewage treatment in South Africa (Pretorius, 1971). This took the dilute sewage water, not settled sludge. The digester (of 2 m^3 capacity) was akin to the sludge-blanket digesters mentioned elsewhere. The sewage water flowed upwards through the cylindrical digester, the flocculating bacteria and sludge particles being prevented from escaping by an inverted-cone baffle. Gas escaped from the apex of the cone and the water flowed over a circular weir to the bottom of the filter tower. Unhydrolysed solids and excess bacterial sludge were removed at intervals from the bottom of the digester. The liquid detention time in the digester was 24 hours and less in the filter. This plant ran at South African ambient temperature; about 20 °C.

Newell *et al.* (1979) described a two-stage process for piggery-waste treatment. The waste from 500 pigs was allowed to stand in a holding tank for five to seven days at ambient temperature. Some digestion of the sludge solids and acidification took place in this tank. The supernatant liquid was passed to a rectangular filter of 2000 UK gallons (9·1 m^3) capacity filled with 1½ in limestone chippings. The filter was constructed from ⅜ in (9·5 mm) steel with reinforcing ties and with a floating top under which gas collected at a pressure of 0·5 psi. The liquid from the holding tank was dispersed in the bottom of the filter by pump pressure through ten 4 in diameter pipes with ¾ in holes every ten inches. The filter was heated to 30 °C, but the method of heating was not described. The liquid retention time was presumably three days, as in the laboratory test unit. A similar 3000 gallon (13 m^3) filter was operated on dairy waste-water at a detention time of 0·75 days and a temperature of 26–27 °C.

THE UPFLOW SLUDGE-BLANKET DIGESTER

The South African digester previously described was a digester of this type, but more recent work done in Holland (summarised by Lettinga *et al.*, 1980) has extended this process to full-size plant. A 50 m^3 digester is treating potato-processing waste-water and a 200 m^3 plant is treating sugar-beet-processing water. As previously discussed this type of digester is best used for low-strength, low-solids wastes of high flow rates and is an extension of the contact-digester process and anaerobic filter.

Rectangular tanks have been used for the full-scale plant. The feed is pumped to inlets in the bottom of the tank, one inlet per $10\,m^2$ area being used in the $200\,m^3$ digester. The digester bacteria grow in a flocculent form and the flocs tend to form a floating mass in the lower part of the tank. The upward flow of feed liquid keeps this sludge bed mobile and flocs which break off into the upper part of the tank and any suspended solids from the feed are trapped under rectangular 'boxes' of an inverted conical section and directed down again into the liquid. Gas collects in the top of these boxes and is led off through pipes. A number of the boxes are arranged side by side across the digester tank. The spaces between the bottom edges of the boxes are partly blocked by a 'sealing strip' which allows liquid to pass, but tends to deflect gas bubbles. The liquid level is above the top of the gas collectors and solids which get past the gas collectors settle out in the top layer of liquid, so that relatively clear water can be collected and allowed to overflow from the top of the digester.

THE FLUIDISED-BED DIGESTER

As mentioned previously this is an extension of the anaerobic filter where surface area of the microbial mass is greatly increased by using a matrix of very small inert particles which remain suspended in an upflowing liquid stream. Jewell (1980) briefly reported large-scale laboratory experiments of himself and colleagues on settled domestic sewage water and synthetic media. Efficient removal of pollutants from these low-strength wastes was achieved at high flow rates with liquid retention times of 30 minutes to a few hours. This type of digester is, obviously, suited only to feedstocks of low solids content such as certain factory wastes, and although Jewell reported the successful treatment of cow slurry diluted to 2% solids, such wastes are not likely to be met with in general agricultural practice. However, a digester such as this, or an upflow filter or sludge-bed digester might be used for secondary 'polishing' of settled liquid effluent from a stirred-tank digester treating high-solids wastes.

DIGESTERS FOR AGRICULTURAL WASTES AND VEGETABLE MATTER

Basically, digesters for this type of feedstock do not differ from the sewage and other digesters previously described. However, as described below, the nature of the feedstocks and the economics involved have imposed different problems on the building and running of the digesters. The solution of these problems has led to a variety of different designs of digester and some of these designs are described below.

Stirred-tank Digesters

The digesters are generally of the single-stage, stirred-tank variety as they have to deal with slurries of between about 3 and 10 % TS, much of which is suspended solids. There are, however, other designs which are for use with the almost solid wastes of 30 % or so TS produced in some animal- and poultry-rearing systems. The digesters described in this section differ, though, from the Gobar and similar digesters previously discussed in that what is aimed at is an automated, heated and stirred digester which will cope with minimum attention with the large volumes of waste produced on modern farms, or with large volumes of vegetable matter, and will produce comparatively large amounts of energy. Many also have to work in a cool or cold climate.

Automated, or semi-automated digesters for farm wastes were first tried many years ago, for instance in Germany during and after the last war, but these plants failed or were discarded for the economic reasons previously mentioned, and it is perhaps fair to say that none were completely successful in a mechanical sense. One of the biggest problems was scum and crust formation. Animal excreta and plant materials form a different kind of surface layer than do sewage sludges, and one much more difficult to control. Sewage sludges tend to form an oily scum which can be broken. However, animal wastes contain fibrous particles of feed residues such as grain husks, grass stems, straw fibres, and also animal hairs. In addition, spilled feed such as silage or dried grass gets into the waste. Fibrous materials such as straws, wood shavings and sawdust, dried corn cob residues, and so on, are often used as bedding for animals or for poultry litter, and these materials get into the wastes. The fibrous materials are only slowly digested and they tend to float to the top of the digesting sludge, in part buoyed up by gas bubbles. Here they can very quickly form a mat which dries out on top and which gradually thickens. The authors and their colleagues in Aberdeen have had experience with their earlier piggery-waste digestion plant of crusts forming in a few weeks which could only be broken up manually. Plant materials added as additional feedstock to animal excreta will increase the fibre content of the slurry, although green plant fragments will tend to be more easily wettable and degraded than the drier straws, etc., used as animal bedding.

Some of these earlier digesters were built by local labour and to local designs and reports on them are difficult to come by. Various tanks, heating equipment (including coal-fired boilers) and stirring and mixing mechanisms were used. Mixing, if done at all, seems to have been usually by hand or machine-operated paddles. A report on a German digester (Anon., 1951) said that a first digester stirred 'as in a municipal plant' was not successful as a crust formed, so the digester was redesigned. The digester had two tanks operated in parallel, the pumped inflow being shared equally between the two tanks. The first plant had two 30 m^3 tanks, a later one two 100 m^3 tanks. All were made of concrete with fixed tops and a separate gas-holder. A single pump with various bypass pipes was used for input, output and mixing. For mixing, thin sludge was pumped out of the lower part of the digester and was returned via a jet pipe which was moved up and down the centre of the tank and at the same time rotated. The means of moving the jet pipe is obscure. Mixing occurred three times a day for 15 minutes each time. The digester contents stratified, though, and a thick sludge of larger particles collected at the top. This thick sludge was apparently removed after two to three weeks running. The feedstock was animal slurries plus vegetable matter and the daily charge seemed to be adjusted to gas production (i.e. if gas production fell off, loading was increased). The digester ran at 25–30 °C and heating was by an external sludge heater which seems to have been operated only every few days and gave a varying temperature. The method of heating water (to 70–80 °C) for the sludge heater is not stated.

The cessation of war-time restrictions and the return of cheap and plentiful fuel stopped further work to solve the problems found in these earlier digesters and more recent work has had to start again from the beginning. But although problems of pollution by animal excreta were and still are some of the reasons for research and development on anaerobic digesters, energy supply has become of increasing concern. While the decrease in pollution and increase in ease of handling of farm animal excreta brought about by anaerobic digestion is of importance to the population as a whole, in only a few cases can an actual monetary value (in terms, say, of penalties if a pollution is not controlled) be put on this aspect of digester operation. With most farm digesters the return on energy production is the main consideration, and, of course, some digesters are now being built purely for energy production. These digesters, then, differ from the sewage-sludge digester where pollution control is the main concern, and gas production is not properly costed. Production costs of the digesters described in this section must be kept down and the solutions found to problems must be cheap as well as efficient. Running costs and energy inputs must also be as low as possible. Mosey (1980) recently quoted costs for a 230 m^3 sewage digester as about £94 000, without gas engine and generator. He concluded that even with the usual working life of a sewage digester being 30–50 years, construction of such a digester could not be contemplated on the value of gas produced alone. To be a viable proposition on a farm the cost of this size of digester would have to be about a third of that quoted, and this price range is being attained. Because of the cost factor, and because many farm digesters are smaller than those commonly used in sewage works, and because the feedstocks differ from sewage sludges, little of the technology used in sewage-digester building can be applied to the agricultural-waste digester. Since building of farm-waste digester plants is only just beginning, some economic models on gas production from animal excreta and vegetation have taken costs of sewage-works digesters and are thus likely to come to an incorrect conclusion. Although the economies of scale predicted by the models do apply, the economic size for the digesters is likely to be much smaller than predicted. But even applying the costings now being obtained to digesters for farm wastes and vegetation there is still a lower limit below which a digester, while viable as a pollution-control system is not viable purely as a producer of energy. Some of the factors involved were discussed by Mills (1979) and in a Swedish review and modelling exercise produced a few years ago (Metangas ur Gödsel, 1976). The absolute values are affected by the type of plant used for feeding and controlling the digester, and the conventional fuel for which the gas is substituted (and this depends to some extent on national taxes, etc.), but at the present time the break-even point for power-production with a stirred-tank digester built to the kind of specifications with steel or concrete described later, is probably represented by the waste produced from about 2000 fattening pigs or 75–100 cattle. Smaller-scale glass fibre and plastic digesters have been built which will provide, say, domestic hot water, but the automated digester for a family and a few pigs or half a dozen cows is not a viable proposition.

At the present time many experimental or full-scale plants have been or are being built, and a variety of shapes and sizes and operational methods are being used. Some of these plants have been built by farmers themselves, some have been built by research teams or by commercial firms, or by the latter acting in concert. All plants have experienced much the same difficulties with crust formation and pipe and pump blockages by the fibrous solids of the wastes. As mentioned before, the only way to overcome the difficulties posed by the feedstocks is to build large-scale plant, test it, and modify it in the light of experience. So the present generation of automated digesters is to some extent developmental. Better digesters will be built in view of present experience.

Some digesters have been working for some time, but it is difficult to determine exactly how successful some of the farm plants are. While digesters will always need some attention, there is a natural tendency on the part of those who have built plants themselves to ignore, or 'gloss-over' as part of normal running, periods of down-time or manual cleaning of scums or blocked pipes, which to the buyer of commercial plant wanting a system that will just 'go' would seem intolerable. Costings, either capital or running, of partly home-made plants are often inaccurate and neglect commercial considerations.

The digester tanks themselves have, in some cases, been modified silage towers or other tanks available on the farm. Such tanks can be used, with saving on costs, and indeed one commercial development of agricultural digesters is based on a 'glass-lined' steel tank which has been sold for a number of years for farm slurry storage. These tanks are corrosion resistant and are built up from standard sections which can give a number of different dimensions and volumes. Roof sections can be supplied to make a fixed-top tank, or a floating top can be built. The tanks are built up on-site from the sections, on a concrete base; such tanks are comparatively cheap. To avoid fracturing the glass lining it is best to have pipe ports, inspection hatches, etc., fabricated before assembly, although using suitable techniques, holes can be cut in an existing tank. Tank tops can be made of steel, as the tank, or of plastic on a metal frame. Such steel digester tanks have been built with capacities from about $200 \, m^3$ to $1300 \, m^3$. The form of the tank can be varied from 'tall' to 'squat'.

Other steel tanks have been used in digesters. The Aberdeen experimental plants (of $13 \, m^3$ capacities) are fabricated in welded mild steel plate (Robertson, *et al.*, 1975). No corrosion problems have been encountered in the inside of the tanks, which have fixed tops, but one which got water under the insulation suffered rusting on the outside over a period of some seven years. The rusting was not, however, sufficient to penetrate or weaken the tank, but caused severe surface pitting and flaking. The mild steel is good for experimental small plant as it can be relatively easily cut and drilled for attachment of new pipes and other equipment, but is probably too expensive for commercial digesters in most cases. However, a 1500 tonne capacity steel tank used for storing palm oil was adapted as an anaerobic digester in Malaysia, and subsequently two fixed-top, steel tanks of 3500 tonnes capacity were specially fabricated as digesters. These latter were built to contain gas pressures of about 48 in WG (Morris, 1980).

Cast concrete has been used for digesters of various sizes, for instance a small commercial digester ($45\,m^3$) known to the authors. Some problems of gas seepage were encountered which were overcome by sealing the interior walls with a resin-based paint, but there was some tendency for this to flake over a year or two's running. Another design used in experimental plant in New Zealand is based on the interlocking concrete panels used for commercial slurry or grain stores. The tank is fitted with a butyl-rubber lining which also forms the top of the digester, where it is held in place by steel bracing. These panels are flat and form a 'multi-sided' tank. The experimental plant is $45\,m^3$ in capacity (Stewart, 1980), and big tanks could not be built on this principle. However, some forms of concrete slab construction giving a water-tight cylindrical tower could be used. The cast-concrete or steel tanks are self-supporting.

Glass-reinforced plastic has been used for some small ($c.\ 10\,m^3$) cylindrical commercial digesters with fixed tops (Chesshire, 1978). However, this material would not be suitable for larger digesters unless structural stability was ensured by a steel framework, and unless a large number of duplicates are to be made the costs of moulded plastic equipment are high.

Another type of digester manufactured in Britain has a concrete tank of rectangular plan with sloping sides. This has a patent floating gas-holder roof made of interlocking rows of concrete T pieces, which provide weight, separated by flat glass-reinforced plastic panels also interlocked with the T pieces (Dodson, 1978). The advantages of this construction are that any area can be covered by adding more T pieces and panels, and that with the legs of the T pieces being submerged in the digesting sludge, the rows divide the digester top into a number of separate gas-collecting compartments. The cover panels to a compartment can be lifted off to inspect the sludge or the working of gas mixers or the heat-exchangers (see later), or for maintenance, without disturbing the whole top or gas collection from the rest of the sludge. The concrete pieces are simple castings and the plastic panels are easily made. However, they have been found to be rather expensive and thin steel panels are now being experimented with. The first digester made to this system has a sludge volume of $227\,m^3$.

Probably the biggest digesters so far made for gas production from animal wastes are two on a very large cattle feedlot in Oklahoma. Each digester is 100 ft in diameter and 36 ft high with a sludge depth of 35 ft, giving a working volume of about 2 million US gallons ($7570\,m^3$) (Meckert, 1978). The digesters are fixed-top, but their mode of construction was not described in the paper referred to.

In Japan a number of digesters of medium size have been constructed at experimental farms. These are to a number of designs (Haga *et al.*, 1979). One, for example, is a rectangular concrete tank built underneath a 30 cow dairy unit of capacity 220 m³. The tank is gravity fed from the channels under the slatted floor of the cow house (Fig. 5.6) and the digested sludge flows out of the riser pipe. Another digester consists of two 31·4 m³ horizontal, cylindrical, steel tanks, above ground and pump-fed with piggery waste. Each tank acts as a separate digester.

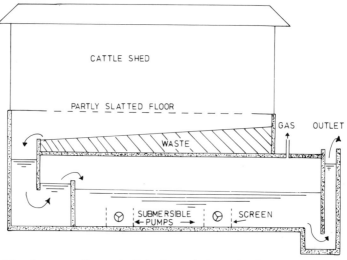

FIG. 5.6.   Japanese digester built under a cattle shed (after Haga *et al.*, 1979).

While there are many digesters, now being made or running, of up to about 50 m³ capacity, the number above this size is much less. Here only a selection has been made to give an idea of the varied types of construction used.

As previously mentioned, concrete has some insulational properties and it is not usual to further insulate a cast concrete digester, although burying the digester undergound or banking earth around it does add to the insulation. The steel tanks do need insulation to increase overall efficiency, although the bigger the tank the less the heat loss per unit volume. This was shown by Mills (1979) who calculated the heat loss from a digester running at 35 °C in an ambient temperature of 0 °C and with a detention time of 10 days. For a 10 m³ digester heat loss was 0·3 W; for a 500 m³ digester heat loss was 4·3 W. This was for a cylindrical tank of height to diameter ratio of 1:1. It has previously been pointed out that this gives the least surface area to volume ratio. Insulation has generally taken the form of plastic 'foam' such as is used for house or chemical-plant insulation. With one Aberdeen digester 100 mm of isocyanate foam gives a theoretical U value of $0.34 \, W/m^2/{}^\circ C$ (Mills, 1979). This foam is applied outside the digester under a thin, welded, sheet-steel skin. The foam was injected between the inner and outer walls of the tank through numerous holes in the outer plating which were subsequently sealed with rubber plugs. The insulation then set to a solid. Glass-fibre lagging under a riveted, galvanised, steel-plate skin was found not to be successful in the first Aberdeen digester, as water was drawn in between the riveted sheets and the insulation became permanently damp, and so useless. Waterproofing of insulation can also be done with various kinds of plastic surfacing of the foam. In one type of commercial digester plastic insulation is applied to the inner surface of the steel tank. Most of the plastic-foam insulations are 'adhesive' and will stick to a well-cleaned, dry, steel surface so mechanical support for the insulation is not necessary.

Cooling rates of about $0.5 \, {}^\circ C$ per day can be obtained with above-ground digesters running at $25\,{}^\circ C$ to $30\,{}^\circ C$ above ambient temperature in a cool climate and in some cases better results may be obtained. However, heat loss can also occur from weir and standpipe overflow systems as these have to be vented to air to prevent syphoning and warm, moist air can be given off through the vent. Weirs and overflow pipes should be lagged and covered as far as possible, and obviously all hot water and sludge recirculation pipes should be lagged.

Heating and mixing of these farm-waste digesters has been done in a variety of ways. The Aberdeen digesters are heated by sludge circulation through an external heat-exchanger. In one, sludge is circulated by a pump of capacity 113 litres/min through the tubes of a tube-in-shell heat-exchanger fed with water at $50\,{}^\circ C$ from a gas-fired boiler. The latter is a commercial domestic hot-water boiler with slightly enlarged gas jets to main and pilot burners. The boiler capacity is 45 000 Btu/h (13 kW) for a digester of $13 \, m^3$ capacity and 10-day detention time running at $35\,{}^\circ C$. The boiler and sludge pump are switched on when required by a thermostat system monitoring digester temperature (Robertson *et al.*, 1975). While high sludge velocities and high water temperature make heat transfer in the sludge-circulation heater efficient, the system requires a relatively high-power pump to circulate sludge and so is costly both in capital and running. Hot-water circulation inside the digester needs only a small water pump and is economically and energetically low in running costs. The only drawback is that the heat-exchangers are not easily accessible in the event of a leak or other failure. The question of keeping internal heat-exchangers free from sludge build-up has been mentioned before.

However, because of the low running costs internal heating is the most favoured system. Some smaller digesters (up to about 50 m³) are heated by water-pipe coils either fitted around a mechanical paddle stirrer or with a free-rising gas-stirrer underneath, to circulate sludge for good heat transfer and to keep the coils clean. Steel or plastic pipes can be used. Copper or brass pipes or fittings should be avoided in digesters as not only do they corrode due to sulphide formation but copper is toxic to digester bacteria, as previously mentioned. This system could be used on bigger digesters. Brade & Noone (1979) described a domestic sewage digester 24 ft (7·3 m) high by 31·5 ft (9·6 m) diameter heated by a central coil with a 5 ft (1·5 m) diameter stirring paddle working underneath the coil. Some farm digesters are heated by flat heat-exchangers in steel, similar to domestic room-heating radiators. The concrete and plastic roof digester previously mentioned (Dodson, 1978) is heated by banks of vertical plastic heat-exchangers. These are of plastic sheet with holes running through, capped at top and bottom by water inlet and outlet pipes. This plastic sheet is sold commercially for solar heat exchangers. Each bank of heat-exchangers is suspended under one of the plastic digester top units and so can be removed separately for inspection or repairs without disturbing the rest of the digester. The hot-water flow is from valved pipes connected to input and return pipes running alongside the top of the digester. Gas jets under the heater banks circulate sludge and keep the heaters clean.

Gas draught-tube heat-exchangers are also used for heating and stirring. An experimental, steel, draught-tube heater mixer built for one of the 13 m³ digesters in Aberdeen consists of five sludge tubes inside a single water tube. The latter is fitted with internal baffles to break up the direct water flow. The mixer is located in the middle of the digester and extends from just above the bottom to below the top of the liquid. A small model system was tested during the design stages of the tube and the heater was designed to transfer the energy from a 13 kW, digester-gas fired, boiler at a gas flow rate of 0·32 m³/min which would give a liquid flow rate of 62 m³/h. The cross-sectional area of the digester was 3·8 m² so the superficial gas velocity in the digester vessel was 0·08 m/min. The heater water temperature was the 50 °C used in the external, sludge-circulation heat-exchanger and the digester-running temperature was 35 °C. Tests with water showed that the design expectations were fulfilled and even with the loss of liquid flow of between 10 and 30 % expected from tests with pig slurries in place of water, mixing and heating should be quite sufficient (Mills & Montgomery, 1979).

Extended tests are in operation. One difficulty found with draught-tube mixing in small digesters is a side-to-side swell over the surface of the liquid which can uncover overflow weirs at intervals. Baffles might break up this if it occurs. A similar system is being built into a commercial 600 m³ piggery-waste digester near Aberdeen.

A digester at a State farm in America being used in developmental work is built from a fixed-top glass-lined steel tank of 189 m³ capacity and uses a central, water-heated draught tube with gas piped in a number of small tubes running down the inside of the draught tube and allowed to bubble freely from the bottom of these tubes up the draught tube. The cattle slurry inlet to the digester is under the draught tube. The draught tube terminates below the digesting sludge level and a deflector plate is fitted at the sludge level to force sludge from the tube sideways and down into a liquid. (Ecotope Group, 1977). As this is a developmental plant, operation of the gas mixer has been varied. A later report (Ecotope Group, 1979) said that gas flow was then successfully being operated at 15 min on/15 min off, with a saving in gas-compressor energy. Possibly even shorter mixing times could be used. Gas flow rates were not given.

Although the heater mentioned above used water at 50 °C, water temperatures of 40–45 °C are often used in internal heaters, particularly the coiled pipe or radiator type. These are in digesters running at about 35 °C. The water temperature can vary with the planned liquid velocity over the heater surfaces, but the sludge in contact with the heater should not be raised to more than about 45 °C or there is danger of killing mesophilic digester bacteria. To prevent thermal shock to the bacteria a temperature rise of not more than 1 °C per hour in the whole digester contents is usually aimed at.

Sunshine on a digester and input tanks can raise the temperature so much that no other heating is required, but in most climates such an occurrence is only occasional. However, in a suitable climate solar heaters might be incorporated in the slurry input lines or in a sludge-recirculation type of heater. Hills & Stephens (1980) have recently reported experiments in California (USA) where a 110 litre dairy-cattle waste digester was run at 35 °C with the input heated by 'bread-box' or 'solar-pond' types of solar heaters. A small amount of supplementary electric heating was needed to keep the digester temperature constant.

A motor driven turbine stirrer has been in use on one Aberdeen digester for some 7–8 years. It consists of two discs with vertical blades, the discs being set on a vertical shaft near the bottom and top of the digesting sludge and being of diameter one third of the digester diameter of 2·43 m. There are also four vertical baffles running up the side of the digester at right angles to the surface and of depth about 15 cm. The bottom bearing is a Tufnol block and half way up there is a Tufnol steady bearing in a frame connected to the sides of the digester. There is a ball bearing in the top of the digester. Gas is prevented from leaking through this by enclosing the shaft in a pipe connected to the digester top and terminating some inches below the sludge level. The stirrer is driven by a 5 kW motor geared down to 50 rpm. Operation of the stirrer was brought down in early tests from continuous to three minutes per hour (Mills, 1977). Less stirring might be possible but it was not tested and the stirrer has been operating at this rate for a number of years. Piggery waste has been digested and the digesting sludge has had a Total Solids content of about 2·5 to 4%, the input sludge varying in different experimental periods with a maximum about 6·5% TS. Other digesters with mechanical, paddle stirrers have been built, but gas stirring is the method of choice, especially for the larger digesters. Some gas stirrers have already been mentioned. An internal gas recirculation mixer was used on one of the Aberdeen digesters (Mills, 1979). The sludge returning from the external heat-exchanger passed through a venturi unit in the head-space of the digester. Sludge moving at 2·5 litres/s drew in 4 litres of gas per second and the gas and sludge mixture was discharged low down near the centre of the digester. The escaping gas formed a free-rising gas mixer which prevented crust formation. This digester has also been mixed by a normal, free-rising gas jet. Gas flow rates for mixing digesters with farm wastes are still a matter for experiment as the sludges differ so much in solids content and size and nature of particles. Flow rates about equal to those quoted for domestic digesters will probably be sufficient with some wastes (e.g. piggery waste of relatively low solids content), but cattle wastes can be of 10 % solids content or perhaps more and the solids are reduced by only some 25–30 % in the digester (see later). Cattle wastes, in the authors' experience, also seem to be more 'gelatinous' than piggery wastes and need more stirring. Gas flows of about 0·1 m³/min/m² of digester area will probably be needed for these (Mills, 1979). Some difficulties have been found with gas mixing in digesters with cattle wastes containing straw. Straw presents one of the most difficult problems in crust formation with farm wastes. Experimentation with frequency of gas stirring, or a combination of gas and a small mechanical stirrer near the top of the sludge has overcome some problems. It is almost impossible to give figures for stirring and mixing. The sludges and the design of digesters and apparatus vary so much that general figures are somewhat meaningless. The apparatus should be designed on general engineering principles, but it will have to be tested under the conditions of use and modifications made to either design or operation (i.e. gas flow rate, time of gas or mechanical mixing) in the light of experience.

Various gas pumps ranging from diaphragm to liquid ring types have been used with farm-waste digesters. The former are quite cheap for the smaller digesters; the capacity and price of the latter tend to rule them out except for the larger digesters.

The various Japanese digesters (Haga *et al.*, 1979) previously mentioned all use submersible pumps in the digesters for stirring and heating. The paper did not mention their performance. A commercial submersible pump made for working in domestic sewage and other materials was tested in some early experiments in Aberdeen. The pump was placed in the bottom of the 13 m^3 digester and used by itself and also with a draught tube fitted to the exit port to lift the sludge to the top of the liquid. Whilst mixing could be attained, power inputs were high and problems, of wear in the pump rotor and casing and sludge caking on the motor which caused over-heating and breakdown, caused the tests to be abandoned.

Mixing by sludge recirculation through the heat-exchanger with a digester contents turnover time of about two hours has not been successful with pig wastes in the Aberdeen digesters, although various sludge inlet and outlet points were tried (Robertson, *et al.*, 1975; Mills, 1979, unpublished). Bulk mixing is adequate but crust formation has not been prevented.

Inputs to the digesters are various. Methods include gravity with a gravity overflow or, in one case at least, a pumped outflow, or by pumped inlet, with outflow over a weir or standpipe system providing the gas seal. Improvements to the Aberdeen weir systems have been made by sloping the bottom of the weir box so as to help any sludge built up to slide back into the digester, and putting a baffle plate below the weir opening to prevent gas bubbles moving up the side of the digester and escaping through the weir, and also putting an internal box to prevent an accidental small fall in liquid level allowing the gas in the digester head-space to escape from the weir. A problem with weirs can be that vigorous stirring can cause 'surges' or 'waves' in the liquid which can momentarily leave the liquid below the outlet slot. Some standpipe systems are less prone to this gas escape. The very large Oklahoma digesters previously mentioned have a gravity overflow into a separate tank which acts as a weir system providing a gas seal for the digester (Meckert, 1978).

For input pumping various types of pumps have been used. The authors and colleagues found that centrifugal pumps, besides being generally of too large capacity for small digesters, had high rates of wear with piggery wastes. The mixture of pig hairs, barley husks and grit can form very abrasive masses on ribs or rotors or in corners of rotor chambers. The steel-scroll-in-flexible-rubber-stator type of sludge pump has been more successful and such pumps have been running for many years on the Aberdeen plants, with repairs, and renewals of rotors or stators, being very infrequent. Two of the Aberdeen pilot-plant digesters (150 litres) are fed by a type of peristaltic pump where the liquid is pushed through three separate heavy rubber tubes arranged as one pipe. The tubes are normally kept closed by compressed air in an outer chamber to each, but are opened and closed by releasing and admitting the compressed air in a sequence which pushes the liquid along the tube. These pumps will deal with solids in the liquid and can be obtained in a number of sizes, but would probably be too expensive and complicated with the compressor and air lines and valves to be used in the usual farm situation.

Whatever the volume pumped, a minimum size of pump and piping is required to pass farm slurries without blockage. A minimum pipe bore of 50 mm is needed for small digesters and pipes of 75 or 100 mm bore are better. Even bigger pipes can be used on large digesters where the slurry volumes pumped ensure a reasonable liquid speed through the larger pipes. Since the flow rate for the input required by a small digester working at 10–20 days detention time is much less than that provided by even the smallest pump which will transfer the high-solids slurries without blockage, intermittent operation has usually to be used. The Aberdeen 13 m^3 digesters are loaded hourly, but the pump 'on' time is only a matter of 1–2 min. As previously shown, the nearer to continuous flow a digester can be worked the better, and hourly operation for a few minutes is better than once-daily operation for a half-hour or so. Intermittent input means that output is intermittent and, because of inertia in the flow of the liquid over the weir, outflow occurs after the input has been pumped in, or after some volume of input has entered the digester (depending on the pumping time).

It can easily be calculated that for an insulated digester with a feedstock coming from tanks at a low ambient temperature (e.g. operating temperature 35 °C, input 10 °C), the heat required to compensate for digester losses is much less than the heat required to raise the input to digester operating temperature. This fact, coupled with greater heat losses from the digester, tends to make running of thermophilic digesters at temperatures around 65 °C uneconomic from an energy point of view unless detention time is much less and gas production is much greater than from a comparable mesophilic digester, or the feedstock is hot (e.g. a factory waste). A successful output–input heat exchanger, as discussed below, could help to reduce the energy inputs to a thermophilic digester.

In the normal digester the outflowing sludge just escapes to cool to ambient temperature in the overflow tanks. If the heat in this outflow could be recovered to heat the input then considerable economies in digester energy inputs could be obtained. Input–output heat-exchangers are not easy to operate for digesters, though. The slurries and sludges do not easily flow through heat-exchanger pipes. Heat transfer rates may be less than for water–water exchanges. And, as just pointed out, the time difference between input and output prevents a direct contra-flow heat-exchanger being used. A coil-in-tank experimental heat-exchanger for one of the Aberdeen 13 m³ digesters and the theory behind this was described by Mills (1979). The digester overflow runs into the insulated tank. The input coil holds one input volume (i.e. the volume moved during the 'on' time of the pump). The input slurry picks up heat from the output in the tank and is then displaced by the next input surge. The sludge in the output tank is displaced by the next overflow. With the input pump set to work every half hour and a 40 mm diameter coil pipe, a heat recovery of 45–48 % of the theoretical was obtained with water in the tank, but operating on the digester, with sludge in the tank, heat exchange fell to as low as 35 % of the theoretical.

The below ground, floating-roof digester previously mentioned has vertical heat-exchangers at each end in the input and output sections, the heat-exchangers being connected by a pumped water flow. It was expected that this system be 60 % efficient (Dodson, 1978). However, practice has shown difficulties and the system was not being used at the time of writing, although the digester was functioning well.

An American digester previously mentioned has an input–output contra-flow heat-exchanger fitted. This consists of the input pipe branching into a number of vertical pipes which rejoin to a single pipe into the digester. Digester output passes into a tank surrounding the vertical pipes and then into a further tank, the whole being a closed system with output being pumped from the digester through the exchanger and tank. The original piping was of aluminium, but this was to be replaced by thin-walled plastic drain pipes. (Ecotope Group, 1977.) The heat-exchanger functioned with low-solids input slurries, but the diaphragm pump used could not move cattle slurries of the desired working consistency (about 10 % TS) through the heat-exchanger and the system was abandoned for some time. A chopper pump installed in the input tank for mixing the slurry by recirculation was also unable to move slurry through the heat-exchanger. However, a scroll-in-flexible-stator type of pump was later installed and first tests showed that this could successfully move even the higher solids slurries through the heat-exchanger (Ecotope Group, 1979).

Input–output heat-exchangers are a field in which more research and development is required, and heat pumps are another possibility of recovering energy from the output. A heat-exchanger could materially help the energy balance of a digester, particularly in cold climates. However, it may not always be needed in practice even there. If an engine-generator is being run on digester gas, then the waste heat from the size of engine needed to generate the electricity required could be more than enough to heat digester and input even on the coldest days. If there is no other use for this heat then there is no point in saving some of it by using an input–output heat-exchanger. If, however, maximum gas production is needed for some purpose and some of the gas is being burnt in a boiler to heat the digester, then a heat-exchanger which lowers the amount of gas burnt is useful. Similarly, where gas is the main product but a small amount of electricity is needed, heat from the engine alone might not be enough to heat the digester on cold days. A heat-exchanger might make the engine heat sufficient at all times.

Experiences with heat-exchangers and slurry handling in general again emphasise difficulties of moving thick slurries of animal excreta and similar materials through piping at relatively low flow rates. There is little reported work on pressure losses in pipes carrying animal excreta slurries, but experimental results described by Howard (1979) are relevant to digester operation. Howard used slurries of pig and dairy cattle manures with the dry matter content adjusted by addition of sawdust and chopped straw, so the results may not quantitatively exactly reproduce some 'natural' slurries of high TS content. Both plastic and steel pipes were used in the tests. A minimum pressure loss with the cow slurry at 7–8 % TS, and at any flow rate, was well marked in the tests with steel pipe but not so obvious with the plastic pipe. A minimum pressure loss with the pig slurry was not found in either type of pipe. However, frictional losses in the pipes are perhaps not so important as bends in pipes and 'rough' constrictions. Large radius bends are reported not to give significant changes in pressure loss compared with straight pipe and experience shows that they do not cause difficulties. However, small radius bends (e.g. right angle 'elbow' fittings) can be the cause of blockages. Another source of blockages is pipe joints where the pipe ends have not been properly mated under an outer sleeve, or a welded or brazed joint left rough. At such points a few animal hairs, fibres or grain husks, or a mixture of these, can become lodged and this nucleus will rapidly build up other fibrous material and slurry particles to form an impermeable mat. This material, even if only a few inches in depth, can resist pressures applied to the following slurry even by relatively high-pressure pumps. In the Aberdeen experimental plants blockages in pipes caused by pig hairs and grain husks have stopped pumps which can apply about 60 psi pressure. Plastic pipes can burst under such conditions. However, proper design and assembly of pipe runs can avoid such blockages.

Straw, however, does pose problems. Long straw has been found to block pipes and pumps as above, but finely chopped, or milled straw has different properties. In the authors' experiments milled straw had to be added manually to a digester running on piggery waste as straw and waste slurries even of 4–5 % TS (with about 1 % straw) would not pass through the pipe lines of the pilot-plant digester (Hobson, 1979). Admittedly this was a small plant, but Pfeffer (1979) also reported that 3 % solids, milled-straw slurry could be pumped only with difficulty and thicker slurries were impossible to handle. Such slurries tend to solidify when pressure is applied.

Another source of difficulty with poultry excreta is feathers. The quills of large feathers can bridge pipe inlets and cause blockages and they can also form surface mats in digesters and block outlets. Feathers can be broken up by maceration, as described below, but since they float on slurried poultry excreta it would be possible to devise some mechanism for straining or skimming them off. In the authors' pilot-plant experiments this is done by hand, but hand skimming would not be feasible on a large scale.

The difficulties with fibrous materials are more apparent with the feedstocks for agricultural-waste digesters. Particle sizes are reduced in the digestion process and the resultant output slurries are much easier to pump and pipe than the raw materials.

To reduce handling problems, feedstocks containing long-fibre particles, or lumps of vegetable matter such as potatoes, need to be chopped to a more uniform and smaller particle size. This can be done in the feed tank to the digester. Even if the flow of feedstock is fairly uniform (say cattle or pig slurry from animal houses constantly occupied) some form of sump will be needed as a ballast tank to prevent the digester input pump from running dry because of an inadvertent stoppage of the flow, or because the pump flow rate over the short periods that it is on is greater than the flow rate of feedstock. Such an input tank may hold only a few hours' input, although one or two days' supply would be more reasonable. The input slurry, if it contains suspended matter, will need to be mixed, or, particularly if digester input is intermittent (as are most), problems of particle settlement or flotation will arise. This can not only result in the digester being fed with slurries of varying composition, which does not aid digester performance, but can also cause pump blockages as the input pump picks up thick sludge from the tank bottom. Feedstock mixing can be done by having a pump which circulates the slurry in the input tank either continuously, or before the input pump switches on. Indeed, it is possible to arrange for the input pump itself to circulate the slurry for mixing before the flow is diverted to the digester. Another system, which has been used for many years with the Aberdeen 13 m³ digesters, is to have a low-speed paddle stirrer in the input tank which is automatically switched on a few minutes before the hourly working of the input pump (Robertson, *et al.*, 1975). In this case a chopper unit can be incorporated in the input pump. In the case of pump-circulation mixing the chopper can be a part of the mixing pump. This is done, for instance, in the American digester plant mentioned above.

If a gravity feed is used for a below-ground digester, some of the problems of handling fibrous feedstocks may be eliminated, although fibres may still build up at pipe or trough bends or behind weirs, and if the flow is slow problems of settlement of solids may be encountered.

The presence of fine grit in wastes from pigs in cement-floored pens has already been mentioned, but while this is abrasive it does not usually clog pumps, and it is carried in suspension through pipes. Adequate mixing in input tank and digester will also keep fine grit in suspension, so that it does not settle out and accumulate in the tanks. One of the Aberdeen 13 m³ digesters is fitted with a large-bore valve and pipe at the centre of the conical bottom so that accumulations of grit may be run or pumped out when necessary. So far the degritting system has not been needed, but this is because the digester has been used over its three to four years of life for a number of purposes, including water-testing of heat-exchangers, etc., and the tank has been emptied and flushed out at times. The other 13 m³ digester has no degritting valve in the shallow conical bottom. This has been running with pig slurry or mixtures of pig and other slurries for the past nine years or so. It has been emptied a few times, at intervals of up to three or four years when some structural alterations have been made, and grit and grain husks have been found in the bottom of the tank, in some cases up to about a foot deep at the edge of the base away from the bottom stirrer disc. However, it is not certain how much of this debris settled out after the stirrer had been stopped and while the tank was being emptied. When an external sludge heater is used, as in these digesters, it is better to take sludge from the top of the digester and return it to the bottom as this will prevent any settled grit clogging the pipes or being sucked into the circulation pump. The only indication of grit accumulation in the 13 m³ digester has been when the heater flow has been taken from the pipe entering the digester a few inches from the bottom, after the digester contents had not been removed for some years. Grit accumulation in digesters will obviously depend on the feedstock used as well as the efficiency of mixing in the tank. The present generation of digesters has not been operative for long enough to really assess the grit build-up. However, it does not look as if it will be a serious problem and most of the larger digesters now being built for agricultural wastes seem to have flat bottoms and no degritting system. Clearing of grit during emptying for routine inspection and maintenance, perhaps every five years, should be adequate. There is no proper way of assessing solids accumulation in working digesters except by determining the effective liquid volume from the dilution factor of known amounts of added radioactive tracers (e.g. ^{82}Br, ^{198}Au) or chemicals (e.g. $LiCl_3$, NaF or dyes) as has been done for domestic digesters (Brade & Noone, 1979; Verhoff *et al.*, 1974; Tenney & Budzin, 1972; Drew & Swanwick, 1962; Burgess *et al.*, 1957).

But while grit may cause few problems, stones in the feedstock can be a nuisance, although this is in the input system as they do not usually reach the digester. The difficulties caused by stones in slurry from earth-floored cattle feedlots which are not found in experiments with slurry from concrete- or brick-paved lots were described by Meckert (1978). Stones jam slurry separators (see later), and they will jam in pumps and pipes. In this American plant (Meckert, 1978) stones and grit are allowed to settle at the bottom of the input mixing tank and removed by a continuously operated drag-chain conveyor. Such a mechanism could only be used on the very largest plants (this one has two 7571 m³ digesters) as the power inputs would be out of all proportion to the energy production of the usual farm or factory digester system. For these digesters a simple form of settling trough in the slurry inflow system seems to be one way of removing stones from the feed, but the exact dimensions and rate of flow of the slurry would be a matter for individual experiment. A settling cone at the bottom of the input tank could be possible, but with this or a settling trough, removal of the stones would have to be non-mechanical to save energy inputs. Some commercial digesters have been fitted with settlement troughs, but in Britain farm wastes mostly come from floored buildings or yards and larger grit and stones are not a continuing problem. Occasionally, of course, a large stone, or a piece of wood or a chain or wire will get dropped into the input slurry, but such accidents and attendant pump or pipe blockages cannot be prevented.

Coarse grit, containing flints up to about 5 mm size and fragments of sea shells up to 115 mm or so long, is often supplied with poultry feeds and this gets into the waste slurry. In the authors' experience with pilot-plant digesters this has not caused a problem, but some grit settles out in the slurry-storage tanks and the remainder can pass through the linear-flow, peristaltic feed pumps used with these digesters (see previously). The grit could possibly cause problems with other types of pump, though, and if it does, then, again, controlled settlement seems the only way of removing it.

One method of overcoming problems of pump and pipe blockages caused by long fibres in feedstocks, particularly cattle slurries, is to remove the fibres with one of the proprietary types of solids separator which are in use for anti-pollution treatment of farm slurries. These separators are of a number of designs with perforated screens which can have holes of different diameters. In some separators the screen is vibrated while the slurry flows on to it. In others the screen is of semicircular or circular construction. Rotating arms alternately pass a roller or a brush over the screen. The roller presses out the fibres on to the mesh, the brush sweeps the fibres away, clearing the mesh. Variations on these designs are also used and descriptions of separators can be found in books on farm-waste treatment (e.g. Hobson & Robertson, 1977).

Separation of solids from slurries will lead to loss of potential gas production. Fibres such as straw may be little degraded in the digester (see later) so loss of gas from removal of these could have only a small overall effect. However, some of the finer particulate matter will be removed with the large particles and this could lead to a more significant loss of biodegradable material. Exactly what will be lost will depend on the type and degree of screening and the feedstock. Experiments on this are being carried out by the authors and their colleagues, but no results are available. Loss of gas production and the energetic and monetary costs of separation will have to be balanced against greater ease of handling of the screened slurry, less downtime with blocked pumps and perhaps smaller input pumps and pipes and less mixing of digester and feed tanks, to determine whether solids separation is worthwhile. Of course, with purely vegetable feedstocks fibre separation will not be possible as much of the feed solids will be fibre, so chopping of the vegetable matter to produce a slurry of small particle size which will pass through pipes and pumps will be the only solution.

'Solid' Feedstocks

Poultry excreta as voided is a solid material with a water content of about 70 %. Cattle excreta scraped from yards or sheds where the urine is allowed to drain away separately may be of 20 % or so Total Solids and will not flow easily. The faeces plus urine slurry from dairy cows is itself of 12–14 % TS. Excreta from cattle or pigs housed with straw, sawdust or other bedding, or poultry deep litter is, again, a solid. Green vegetable matter is about 20 % solids. If these materials are to be used as feed for a stirred-tank digester then they will have to be made into slurries. Experience suggests that a slurry of about 10 % TS is the maximum thickness that can be pumped and piped even if the particle size is small, and 7–8 % TS may be the maximum which can be handled by the smaller pumps and pipelines. Besides mechanical reasons, there can be biochemical reasons (see later) why slurries of a maximum of 6–7 % TS should be used as digester feedstock. To obtain slurries of these consistencies from the materials mentioned above means that they must be mixed into water, and for proper digester operation and uniform gas production this must be done in a regular manner. The consistency of the solid material, whether animal excreta or vegetable matter, should be reasonably constant. So initial experiments should suffice to set the proportion of water to solid necessary for production of a slurry of the desired solids content. A system whereby known volumes or weights of solid feed and water are mixed into the digester feed tank is then required. It seems that this would, in general, be done manually with batches of a few days feed being prepared at a time, as equipment which could do the job automatically would be difficult to make and operate with farm slurries and would be expensive and power consuming. An automatic or semi-automatic system could be possible on the very biggest digesters and such a system is being tested in America to dilute cattle manure to a suitable consistency for processing (Meckert, 1978). Two tanks which were alternately filled to a predetermined level with the solid material and then topped up with a specific volume of water would seem simple to operate manually. To produce a uniform slurry, some form of mixing by a circulating pump or a stirrer, possibly combined with chopping, would then be needed. Chopper–blenders of various kinds are commercially available, and chopper pumps and methods of mixing have already been mentioned.

These methods will produce a slurry that can be digested in a standard, stirred-tank digester, but the digester for solid material has previously been mentioned. The very large, solid-state digester is still only theoretical, but some small units have been built. The solid-state digestion is always a batch process and so a continuous inlet and outlet system is not needed. Vegetable matter could be loaded into large towers in the same way that grass or other vegetation is now loaded into silage towers. That is by blowing the finely chopped material along pipes to the top of the tower or using a belt conveyor. In some cases an archimedian screw feed might be possible. After digestion had ceased the tower contents would be removed like silage from a large opening in the bottom. In a French system at the National Institute of Agricultural Research, Jouy en Jasas (quoted by Wheatley, 1978) animal excreta and bedding of about 14% solids content is loaded into baskets which are then put into a digester tank. The French plants are more for the production of stabilised and odour-free solid manures than for gas production, and this latter in some cases is low. However, the mechanisation required for such plants seems to be high. Digesters can be horizontal or vertical tanks, but require a full-area entry door at the top or one end of the tank. The door has to be provided with a gas-tight sealing gasket and a centre capstan with radial bar closures, or peripheral hand wheels or toggles to close the door firmly. If the digester is horizontal, then the basket of feedstock has to be run into the digester. So some form of bogie and railway line from the loading bay into the digester is needed. A vertical tank, with a top opening, needs a crane to lift in the basket and unload it.

All this, together with the mechanism for loading the feedstock into the basket is costly and consumes energy. Since as previously pointed out, a batch digester plant needs a number of digesters, running out of phase, to provide a continuous supply of gas or to deal with a continuous flow of feedstock, all the loading and unloading equipment, as well as the digesters, would have to be replicated. Or a 'straddle' crane running over a number of vertical digesters and able to carry the charging baskets above them (as has been used) does away with some duplication of equipment. But this would be a very expensive item for a series of large digesters.

The Tubular Digester

In some early work on digesters in Africa, Fry (1974) used a fórm of 'tubular', or 'tunnel' digester. This was on a comparatively small scale and for a pig farm, and the digester was hand operated. Pig slurry was fed into one end of a concrete tube (of circular or rectangular cross-section) which could be heated if necessary by hot-water pipes bedded in the floor. The sludge flowed to an outlet pipe at the other end. The detention time suggested was 35–40 days. The inlet and outlet pipes were arranged so as to give a liquid seal to preserve a gas pressure of a few inches WG and gas was collected in a water-sealed gas-holder. Scum was brought to the top by the natural stirring of the evolving gas and this caused problems. The problem was overcome by removing the scum at intervals with a 'rake' attached to guide rods below the top of the digester. The rake, which cut through the surface layer of the liquid, scraped the sum from one end of the digester to the other where it was pushed out through a 'scum-door' in the end of the

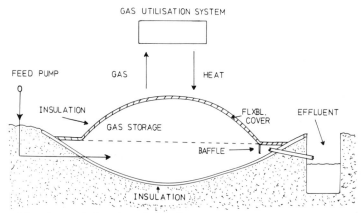

FIG. 5.7.   A practical-scale, tubular digester for cattle waste (after Hayes *et al.*, 1980). The flexible cover, of rubber, acts as a gas-holder.

tank. The rake was moved manually by cables running through the digester.

Later proposals were made for some large digesters to be built in America to this design, but so far as is known this was never done. However, the idea has more recently been taken up again in America where a large pilot-plant (38 m³) to treat cattle waste has been built. This is described by Hayes *et al.* (1980). The digester consists of a concrete trough capped by a butyl-bag top which acts as cover and gas-holder. A side view is shown in Fig. 5.7. The trough and top are well insulated and the digester is heated. The trough walls are insulated with 10 cm of foam glass and the top with 9 cm of fibreglass. The liquid at the ends of the trough is covered with 5-cm polystyrene slabs. The earth under the digester is drained to keep it dry and so improve insulation.

A pump feed at one end of the digester is used, with a gravity overflow at the other end. Although there is provision for recycling it would appear that this has not been used.

Whether these digesters are behaving as true plug-flow tubular digesters is perhaps doubtful. Various aspects of this have been discussed in Chapter 4.

GAS-HOLDERS AND GAS HANDLING

The gas from sewage digesters is almost entirely used at low pressure in boilers or engines, and it is used as it is produced. In some works a small proportion of the gas is dried, scrubbed and compressed into high pressure cylinders (about 2000 psi), but this is sold at comparatively high price to specialist gas suppliers for sale to laboratories and others needing pure methane in small amounts. Only in this way can the cost of cleaning and compressing the methane be recouped and the energy used is only a small proportion of the total gas energy available.

For general use on factory or farm, gas at pressures of a few inches WG is required and the mixture of methane and carbon dioxide can be used as it is. This does mean that the gas must be used almost continuously as storage of large amounts of gas at low pressure is economically unrealistic in the farm or factory situation. With digesters producing one to two digester volumes of gas a day storage of more than part of a day's supply is generally the most that can be contemplated. In practice, storage varies from two or three to perhaps twelve hours' production.

The use of digesters with floating tops for low-pressure gas storage has already been mentioned, and such a system can be built on any scale. If the digester has a fixed top, then a separate gas-holder is needed to store the gas and regulate the pressure. The water-sealed gas-holder, as was used in the old town-gas plants, provides better pressure regulation, is more robust and is inherently safer than other types of holder. These gas-holders are usually built of steel. They are expensive, but will last for very many years. A small, commercial, water-sealed gas-holder is produced in glass-reinforced plastic. The water tank can be given extra rigidity by being sunk in the ground. Corrosion is non-existent, but gas-holders could not easily be built as big as those in steel, and if built would probably be as expensive. The sealing water acts to some extent as a gas scrubber and it might be better drained off and renewed every few years as sulphide sludges and other contaminants collect in it. The water also acts as a safety valve if production exceeds storage capacity or some blockage occurs in a gas line as excess pressure is released by the gas bubbling out through the water seal. (Gas can escape from a floating-top digester in the same way). In hot climates the water may need topping up to replace evaporative losses, but increases due to rain will flow out through overflow pipes. The greatest difficulty occurs in cold climates where the water may freeze. In the authors' pilot-plant the gas-holder has a water capacity of about $1 \cdot 3\,m^3$ and this is protected to about $-5°$ to $10\,°C$ by using a mixture of water and motor-car antifreeze (glycol). The bigger gas-holder (water capacity about $9\,m^3$) serving the $13\,m^3$ digester is protected by using a water–oil emulsion as filling and by heater tapes wound around the outside of the water tank under a glass-fibre insulation with a thin sheet-metal cladding. The heater tapes are switched on automatically as the ambient temperature falls to about 0 °C or just below. Temperatures of − 15 °C or lower, occasionally experienced, have overcome all these precautions, however. Using a water–glycol mixture or heaters is expensive for large installations and for protecting against very low ambient temperatures. Where very low temperatures are regularly expected or prolonged then the expense may be necessary. However, since the gas-holder is small compared with the digester such expense may not be a large part of the total capital or running costs of the plant. If there is excess hot water from engines or boilers, as previously mentioned, then this could be used to warm gas-holder water during winter. Partial burying of the gas-holder may help to prevent freezing, as may enclosing it in some form of building; although in the latter case suitable safety precautions need to be taken, and these, such as adequate ventilation might negate the whole purpose of the building.

A digester top made of rubberised fabric, or butyl rubber alone has been used in some cases as a gas-holder, but this, unless it is filled only to take up the slack in a loose 'bag', will add a varying pressure to the gas. A framework to support the collapsed top is required, and even if the top is not stretched and pressure control is by a weir or similar device on the digester overflow, the flexible top does not seem capable of giving as good performance as a rigid, floating top. In case of an overflow blockage the flexible top does not act as a safety valve as does the floating top and so some other form of pressure release is necessary. If the top is made to expand, then a pressure regulator of some kind in the gas line will be required to give a constant line pressure. Some cylindrical tank digesters have been built with this kind of top, but they have not been in regular use long enough for their durability to be assessed. One point of weakness could be the fastening of the fabric to the top of the digester tank.

Butyl rubber, which is impervious to gases, has been used or proposed in various forms of 'bag' gas-holders. But, so far as is known these have all been small units of a few cubic metres. They are varied in design. Some are unsupported bags, others are contained in frameworks or a walled enclosure. To overcome the problems of varying pressure from an inflating bag, weights, placed either directly on top of the bag or operating through a system of levers, are used. A number of designs have been reviewed by Dohne (1980). One for a 45 m³ digester has been working in Britain for some years. This has a butyl bag inside a wall enclosure, with lever balance weights.

The butyl-bag gas-holder does not seem a really practical proposition for a large digester, and butyl-rubber bags are not cheap, particularly when moulded to a specific shape and made as a 'one-off' or in small numbers. For instance, Dohne quoted costs of low-pressure, water-sealed gas-holders of 100–5000 m³ capacity as 1000 to 120 DM/m³. There is, like the digesters, an advantage of scale. Gas bags with a weight on top of 25–800 m³ capacity cost 2000–25 DM/m³, and simple rubber bags of capacity 5–300 m³ were 600–100 DM/m³. The material is tough, but is more easily damaged than a water gas-holder, and while there is no water to freeze, a black bag will get hotter in sunlight than a water gas-holder, with consequent changes in gas density and volume.

For most digester purposes a water-sealed gas-holder would seem best, and these gas-holders store gas at about 6 in (15 cm) to perhaps 10 in (25 cm) WG. This is the same pressure range as that given by a floating-top digester. Storage of gas at higher pressures means that some form of compressor must be used. Although gas formation by the bacteria is not affected by high pressure, building a large digester in which the gas was produced under pressure would be not only costly but dangerous. Some years ago a steel-tank digester in which gas was produced and stored under pressure was designed and built on a small scale, but nothing more seems to have been heard of this.

Storage of gas under pressures of up to about 100–200 psi in commercially available tanks is possible. Tanks taking such pressures are relatively cheap for the volume of gas which they will contain. (Dohne (1980), for instance, quoted 750–300 DM/m³ for tanks of 10–100 m³ volume.) But although the cost of the tanks may be fairly low, the cost of the compressor and of spares for the compressor must be taken into consideration, as also must the monetary costs of running the compressor and attending to it. The energetic costs of the compressor should also be set against the energy production of the digester. However, rules for inspection and testing of pressure vessels laid down by government order in some, if not most countries, or insisted on by insurance companies, must be observed. These will involve pressure testing and inspection by trained engineers. External corrosion, particularly around welds or flanged and bolted or screwed-in pipes, must be prevented by regular paint or other treatment. Internal corrosion could be caused by traces of hydrogen sulphide or ammonia in the gas, or even by the carbon dioxide under pressure, if the gas is wet, although the anaerobic atmosphere will tend to inhibit some forms of corrosion of steel tanks. So although biogas can be stored as it is produced, for pressure storage it would probably be best to dry the gas after scrubbing sulphides and ammonia from it. Compressors and gas lines (which of course, should contain no copper or brass with a gas containing hydrogen sulphide) must also be kept clean and corrosion-free and regularly inspected. Since the tank is sealed, safety devices must be fitted. The compressor system will have to be automatically controlled. The controls should guard against over-pressurisation of the gas as is normal in pressurised systems. However, since the compressor will be drawing gas from a digester whose gas production rate will fluctuate, the system will have to control against 'over-suction'. If gas is drawn off at a rate greater than it is produced, the various results of this, whatever the design of digester, probably do not need to be described. The provision of over-pressure safety valves suitable for use in an atmosphere that can contain corrosive gases is a problem. While any normally-closed safety device should be tested at intervals (perhaps a few months) by manually (or preferably by internal pressure) causing it to 'blow', this cannot completely guard against failure to work at a critical moment, caused by corrosion. Some form of bursting-disc, if it can be arranged, is probably the safest safety device. Finally, the pressurised tank has to be fitted with a pressure-reducing valve in the outlet gas line to reduce the gas pressure to the few inches WG needed for operation of boilers, engines or gas lights.

While storage at medium high pressures cannot really be recommended for most digester installations, and by far the majority of digesters now in use have low-pressure gas storage, the storage of methane at very high pressures is definitely not recommended for any but very specialised uses such as the sale of pure methane previously mentioned in connection with sewage digesters. In this case the plant is under control of trained personnel and is properly inspected and maintained. And the monetary value of the specialised use compensates for the cost of the compression plant. For high-pressure storage the gas is compressed to 2000 psi or more. Gas under this pressure must be kept in small (a few cubic feet) capacity, very heavy, cylinders of the type used for small quantities of nitrogen and other commercial compressed gases. These cylinders have to be pressure-tested every few months in Britain, if not elsewhere. Compressors and ancillary piping must also resist the same high pressures, as must valves and gauges. Corrosion problems found at lower pressures can be increased at high pressures, and the high pressure storage of scrubbed and dried methane alone will rule out corrosion problems.

Pure methane is stored as a liquid in natural-gas tankers and underground reservoirs, but storage of liquefied gas on a small scale is impossible. Methane will liquefy only below about $-164\,^{\circ}C$ and with digester gas the carbon dioxide, which solidifies at $-78\,^{\circ}C$ would create problems, as would trace impurities.

Storage of digester gas or methane from the gas by absorption or solution has been suggested. So far as the authors are aware there is no solid absorbent for methane. The solution of methane in a liquid hydrocarbon such as butane has been suggested, but Picken (1978) reported that his group had experimented with almost every hydrocarbon possible without success. Not enough methane could be dissolved to increase the energy value of the hydrocarbon more than about 2%. Dohne (1980) also pointed out that solution of methane in liquefied hydrocarbons required refrigeration and this ruled it out for farm situations.

While mentioning liquefied hydrocarbons it might be pointed out that it is a common mistake to equate methane with its higher homologues propane and butane (LPG) and assume that it can be liquefied at climatic temperature as can LPG.

The volume of gas produced must be measured, not only to monitor use of the gas, but also to monitor digester performance. Although the volume of gas coming off a digester will vary slightly from day to day because of complexities introduced by the metabolism of a heterogeneous feedstock by a heterogeneous population of bacteria, over the longer term gas production should be uniform. Wet and dry meters have been used in the Aberdeen plants to register unscrubbed gas. The water in wet meters should be changed at intervals (its condition can be seen in the level glass), and both types will eventually corrode and become erratic in operation or stop, but this should take some years. The dry meters are cheapest to replace. A float flow-gauge with a sensor for the float position proved unworkable in practice as gas flow rate varied almost from minute to minute. Even in the authors' pilot-plants, which are continuously stirred and loaded every few minutes, gas comes off in short bursts. In the $13\,m^3$ plant a large proportion of the gas comes off in a burst while the stirrer makes its once-hourly operation (Mills, 1979). The composition of the gas coming off during operation of the stirrer and between the operations is, however, the same.

Gas Purification

The gas coming off a digester will be largely a mixture of methane and carbon dioxide. If there are any small air leaks into the digester the gas could also contain a few percent of nitrogen. For the reasons previously discussed, gas from the digester should not contain oxygen and if the gas does have a small oxygen content this will denote a leak of air into some part of the system beyond the digester itself. However, such leaks are unlikely, as in most systems the gas is under slight pressure in the gas-holder and this will let gas leak out rather than air in. A leak on the suction side of a gas compressor might introduce air. The gas contains some trace impurities. The main ones are hydrogen sulphide and ammonia, but very small traces of acetic acid, and possibly higher volatile fatty acids, sulphides, amides and other vapours may also be present.

The gas will also contain water vapour and droplets and very small particles of solids as it comes from the top of the digester. In our own plants a trap in the gas line a few inches above the top of the digester returns any spray and solid particles into the digester. The trap is just a wider piece of tube with a baffle inside of inverted-cone shape which interrupts the direct flow of gas. The trap is made from iron piping, as are the main gas pipes. The gas passing the trap then contains water vapour and as the gas is warm from the digester, water vapour will condense out as the gas cools in its passage through pipes. Condensate pots are needed in strategic places along gas-pipe runs, particularly at the bottom of down pipes. In warm and sunny weather the top of a digester and gas pipes in air can be quite hot so water vapour can be carried quite a distance until the gas pipe enters a cool building or an underground pipe run, when it will condense out. Pots should be placed so as to run off such condensate. On the usual low-pressure digester system only six or seven inches of water in a gas riser pipe will blow-back gas from the digester overflow or gas-holder. Condensate pots can be bought commercially or fabricated and are of a capacity suited to the gas flows being handled. For a small plant two or three litres will suffice. The gas enters through a top pipe and exits through a pipe near the top on one side or even in the top. Condensate collecting below the gas pipe levels is run off through a suitable tap, or pumped out by a hand pump at appropriate intervals. Since a condensate pot acts to some extent as a gas scrubber the water from the pots is usually very foul-smelling, as the hydrogen sulphide and other odorous trace impurities previously mentioned will collect in the pots.

Virtually all the water vapour can be removed from the gas by refrigeration if a dry gas is needed for compression or other special purpose. Molecular sieves and zeolites could also be used as solid absorbents of water and these could be regenerated.

For most purposes the gas can be used after passing through the condensate pots and gas-holder. The gases from digesting animal excreta and vegetable matter do not contain sufficient hydrogen sulphide to warrant scrubbing before the gas is used in engines, and no town sewage works seem to scrub sulphide from the gas. Picken (1978) pointed out that modern engine lubricating oils can withstand quite a sulphur concentration before problems would occur, and that it was unlikely that there would be serious build up before a routine oil change took place. A figure of not more than about 1 % hydrogen sulphide has been quoted at times as a limit to avoid corrosion in engines and this figure appears to be higher than the hydrogen sulphide content of gases from most of the digester feedstocks so far used. Gas from a wide range of land vegetation was found by Badger *et al.* (1979) to be very low in hydrogen sulphide (less than 0·001 %). However, some aquatic energy crops such as seaweeds might be different, as sea water contains appreciable amounts of sulphate as do some factory wastes, such as the mine water previously mentioned. However, it should be pointed out that gas compressors have a lower tolerance for hydrogen sulphide than the internal combustion engine.

Removal of hydrogen sulphide is not generally necessary for burning digester gas. The boiler burners on the Aberdeen digesters have built up a little 'dust' over long periods, but no serious effects of sulphide or corrosion from sulphurous flue gases have been noted.

If hydrogen sulphide is to be removed from digester gas two or three methods are theoretically possible—'theoretically possible', because although all the methods have been used for removal of sulphide from the old 'town gas' or natural gas, it may be difficult or impossible to use them on a low pressure system such as a farm biogas plant where gas is produced and stored at a few inches WG pressure. In such a system the back pressure produced by even a thin layer of solid absorbent which is getting damp or in which chemical changes are occurring may be more than the equilibrium pressure of the digester. And a pressure of a few inches WG will obviously not force gas through a tall column of liquid absorbent. A falling-liquid spray in an absorbing tower offers least pressure resistance,' but this will involve pumped recirculation of the absorbent and possibly filtration of the liquid, unless a purely water spray is used. Water could be used for scrubbing digester gas and would remove some impurities and also some of the carbon dioxide. However, chemical scrubbing of digester gas has generally been mentioned in connection with plants where for some reason or other the gas is pressurised to at least a few pounds per square inch. Gas scrubbers are best placed before the gas enters the gas-holder as this latter will equilibrate any pressure changes caused by passage through the scrubbers, and the scrubbing will reduce corrosion in the gas-holder.

Hydrogen sulphide was removed from 'town gas' by passing the gas through iron oxide. The oxide was prepared for the purpose and could be spread on some support medium to increase surface area and permeability. The oxide was placed in boxes arranged in series and to regenerate the oxide about 5 % air was added to the gas stream. Reaction with hydrogen sulphide produced ferric sulphide which in turn was more slowly oxidised by oxygen in the air to regenerate the ferric oxide and free sulphur. Water was also generated in the reactions. Most of the sulphide was removed in the first box in the series and after a predetermined time of use the gas flow was altered so that a later box became the first and the ferric sulphide in the (initially) first box could regenerate in the low-sulphide gas passing out of the first boxes. The oxide was finally removed when the sulphur weight was about equal to the oxide weight.

In a well-run system with gas passing downwards through an oxide bed of 12–18 in thickness and packed at a density of 70 lb/ft³ a pressure drop of 2–4 in WG could be obtained, but in a poorly-run system with large amounts of spent oxide accumulating the pressure drop might be as high as 20–30 in WG (H. Crompton, 1978, personal communication). So, there is obviously, as indicated before, little margin for error available in trying to run an oxide purification-system in a low-pressure gas line, and the characteristics of the oxide bed will change as the reactions proceed. The details of design and running of an oxide purification-system will depend on the sulphide content of the gas and the flow rates. A possible size is about 400 g oxide/m³/h gas flow rate.

With a small plant it would probably not be worthwhile regenerating the oxide by oxidation by air in the gas stream or a separate aeration of spent oxide. The used oxide could just be thrown away, but as this, like the oxide–sulphide–sulphur mixture, is noxious it would have to be dumped in some approved place. In previous times there was a market for sulphur from the oxide boxes, but whether this exists now (and for small quantities) is doubtful.

Liquid absorbents for hydrogen sulphide can be used either as sprays or bubble columns. Zinc acetate or some other zinc salts or a simple alkali solution such as sodium carbonate are suitable. A patented commercial process uses a solution of sodium carbonate and bicarbonate, with sodium metavanadate and anthraquinone-2:7-disulphonic acids. While liquid scrubbing has some advantages, solutions such as zinc acetate cannot be regenerated and the precipitated zinc sulphide must be removed and disposed of, and, again, this is a noxious substance.

There is need to remove carbon dioxide from the digester gas only if it is to be used for a particular purpose as pure methane; for instance in a very large biogas plant designed to supply methane to a national gas grid. The capital and runnning costs may then be justified.

Carbon dioxide, being an acid gas, can be absorbed into any alkaline solution, such as sodium or calcium hydroxide. Carbonates and bicarbonates of various solubilities are formed and these have to be removed, and the alkaline solution replaced as it is used up. Monoethanol-amine is used to scrub carbon dioxide and hydrogen sulphide in a plant where large volumes of methane are destined for national natural-gas pipelines (Meckert, 1978).

The traces of ammonia in biogas could be removed by scrubbing with an acid solution, but would probably be washed out during sulphide or carbon dioxide removal.

Uses of Digester Gas

The removal of carbon dioxide from digester gas and the sale of the methane for addition to city natural-gas supplies is possible only on the largest production scale. Except for the one case described later small or medium (say to about $1500\,m^3$ volume) digesters are being built to utilise waste or crops on a particular site and to use the biogas on that site or its immediate neighbourhood, and to use the gas at low pressure. In these cases use of the gas directly or via conversion to electricity is contemplated.

It perhaps seems obvious to say that on-site use of gas depends on having a use for it! But this may involve planning and reorganisation of the running of the various activities on the site.

It has already been suggested that storage of biogas in large volume is not feasible for a number of reasons (including safety). The normal plant could probably store up to about half a day's production to even out gas flow and act as a ballast tank for running of engines or boilers. Such storage could help to meet a peak power demand. If a small site had a demand for, say, 20 kW of electricity over an eight hour working day but hourly gas production could not meet this demand, then storage of gas for the rest of the 24 hours might enable a generator to be run for eight hours at this loading. Digester heating, if by engine cooling water, could be geared to suit this, as could digester loading: the digester being slightly (a degree or so) 'overheated' during this period and being allowed to cool during the rest of the time. With previously quoted cooling rates of 0.5 to $1\,°C$ a day the temperature variation in the digester would be small. On the other hand if the digester were capable of producing a continuous 20 kW output then some other use would have to be found for the energy.

It is permissible, in some countries at least, to sell surplus electricity to public utilities, but at a low price. It is also doubtful if these bodies would really want 'off-peak' power as they usually have a surplus capacity at these times, and there can be technical problems in running power from a number of small generators into a main power line.

The best way of getting over such problems is, where possible, to plan the use of the power so as to even out the load. This is being done with some farm systems. For instance the day could be planned so that milking machines, feed mixers and conveyors, slurry scrapers, and so on all ran at different hours of the day. On the other hand, a dual system might be best with part of the gas being used in a generator (which could also heat the digester) and the rest used in a boiler for provision of heating and/or washing water.

The possibilities obviously have to be planned for each particular site and digester. But there is still one area of debate and that is the provision of standby mains services. Public electricity services may be reluctant to install, or keep in service, power lines and switch gear suitable for the total load on a site, when this current is only going to be purchased in the event of a digester failure—at least without 'rental' charges commensurate with the cost of this total load. This problem is coming more to the fore as digesters come into use, but it seems probable that in the light of overall energy savings customers will be encouraged, rather than discouraged, in provision of alternative energy sources. The other way of overcoming this problem is to have a local standby in a diesel generator or an engine that can use an alternative fuel, and oil or LPG for boilers.

From an economic point of view and also for versatility in use, electricity generation seems at present the best option, although economic comparisons are continually changing. For instance, the Government-controlled, projected increases in natural gas prices in Britain will considerably alter the economic relationships with biogas. However, farm digesters presently installed in Britain all have electricity generation as a main, or sole, aim.

A number of firms make engines for use with biogas, and more intend to do so. These engines fall into two categories, the dual-fuel and the spark-ignition engine, and so far they are falling into two power ranges. The spark-ignition engine is used for the smaller units of about 10 to 75 or 100 kW power output; the dual-fuel system is used on larger units.

Dual-fuel engines have been used for the large (about 1000 h.p. or 700 kW range) engines in sewage-works power stations for many years. The engines run by compression ignition, the diesel oil injection (about 5 % of the fuel energy) being used to ignite the gaseous fuel. Pure methane requires 9·5 volumes of air for full combustion, so an air:gas ratio of about 7:1 is required for the normal digester gas of 65–70 % methane. The engines, being of high compression ratio, are more efficient than spark-ignition engines and the higher efficiency offsets the cost of the diesel fuel. Diesel engines are more massively built than spark-ignition engines, and with their slow running speed (about 900 rpm or less) the large dual-fuel engines have a long life. Running on gas the engines can take a longer period between overhauls. Such advantages of the diesel engine are already suggested as to an extent offsetting the higher capital cost of the small diesel engines now being fitted to cars, and small dual-fuel engines will no doubt soon be available for stationary use.

The small spark-ignition engines are essentially car engines with hardened valves and seats with carburation modified for gas instead of petrol but in some cases a small diesel engine has been used as a base because of its more robust construction. The compression ratio is usually in the higher petrol-engine range, about 9:1. The advantages of biogas as a fuel are that it is of high energy content (about $6·7\,kWh/m^3$ for the usual digester gas, but this varies with methane content), it has good anti-knock properties and it is clean. These engines run at relatively high speed (about 2500 rpm) and questions have been raised as to the life between major overhauls of a small spark-ignition engine running constantly, or nearly so. Opinions vary, and it seems that only practical on-site testing will determine whether a year or a few months is the right order.

The compression ignition engine can run entirely on diesel oil if the need arises. Experiments at Leicester (D. J. Picken, 1979, private communication; Soliman, 1978) have been concerned with the development of a dual carburettor system for spark-ignition engines whereby supply failure in digester gas can be compensated by an increased input of another fuel such as kerosene ('paraffin'). The proportionation can be made automatic from a level indicator on the gas holder, or the supply pressure of gas. Kerosene is a poor fuel which is greatly improved by admixture with methane.

So far as digester gas itself is concerned, experiments (e.g. Picken, 1979) have shown that power output at constant rpm decreases slowly, but the engine performs satisfactorily, as the percentage of carbon dioxide in the gas increases, up to about 45 % carbon dioxide. Above 45–50 % (i.e. above that found in digester gases) combustion becomes irregular with rapidly increasing amounts of unburnt hydrocarbon in the exhaust. A minimum amount of unburnt gas was found at about 40 % carbon dioxide, but the change from 0 % to about 40 % carbon dioxide was not great.

Neyeloff & Gunkel (1975) previously carried out experiments with various mixtures of methane and carbon dioxide in an engine and found that digester gas could be used to fuel a spark-ignition engine. They found that the power output obtained depended very much on the engine design, as well as the carbon dioxide content of the gas.

Gaseous fuels for engines have a high octane rating and so the combustion efficiency can be increased by increasing the compression ratio over that used with petrol. The gas also mixes thoroughly with air and gives a mixture that burns more completely than petrol–air mixtures, and the gas also gives less carbon deposits. Advantages of gaseous fuels are summarised (ONAN Tech. Bull. T-015, 1974) as: excellent anti-knock properties; small amount of contaminating residue; less sludge in oil; no wash down of cylinder lubrication during starting; no tetra-ethyl lead to foul spark plugs and other engine parts; a nearly homogeneous mixture in the cylinder; and less valve burning.

It might be advisable in some circumstances to consider the use of two or more small engine-generator sets rather than one big one. A diminished supply of digester gas might run one small engine at full power, but might not run a large engine at fractional power output. Also, small units can be brought in and out of use as power demands fluctuate.

A low gas pressure, such as from a water-sealed gas-holder (6–8 in WG; 15–20 mbar) is suitable for gas or dual-fuel engines. For an electricity output of 10 kW about $5 \cdot 7 \, m^3$ of gas of 70 % methane content is needed per hour. Two to three times this energy would be also recoverable as heat.

The hot water from heat-exchangers on a gas engine should be enough, or more than enough, to keep a digester at mesophilic temperature. If the engine water is more than sufficient to keep a digester at 35 °C, then some advantage in extra gas production might be obtained in running the digester at about 42 °C (see later). This assumes, of course, that there is no other use for the hot water.

The boilers used for heating water with digester gas are natural gas systems with modified jets. Slightly enlarging the jets will be sufficient for most digester gases, but tests may have to be made. Gas pressures of only a few inches WG (2–4) are needed for running ordinary boilers, so pressure-reducing valves will be needed in gas lines if the digester gas is stored under any pressure. As mentioned before an air:gas ratio of about 7:1 is required (according to methane content) for full combustion of digester gas. Care may have to be taken with pilot jets particularly, as owing to the low flame-propagation speed of methane (43 cm/s) which is reduced by the carbon dioxide in the gas, there is the possibility of blowing the flame off the jet. The hot water circuit from a boiler can be arranged to heat both digester and buildings or two separate boilers might be used. The actual layout obviously depends on the site situation, and many possibilities of link-up with other heat-producing systems could exist. For instance, one commercial digester boiler is linked with a heat-exchanger on an incinerator. A gas boiler and gas engine can also be linked as heat producers.

It may be possible to use a dual-fuel boiler with standby LPG or oil burner to ensure against digester gas failure. Or a separate, standby, boiler may be used, as on the authors' digesters where an oil boiler of similar capacity to the gas boiler is used when, during experiments, the digester gas production is too low.

The efficiency of a gas-boiler water heater can vary from about 60 to 80 % of the fuel energy. Thus some of the modern gas engines with engine-heat-exchangers can be more efficient overall than the best boiler and of the same order of efficiency purely as water heaters without counting the 25 % of fuel energy converted to electricity.

A modern, flash-steam boiler might also be used to run a steam engine or steam turbine, but so far as is known this has not been tried. However, gas turbines have been tested at some of the large sewage works. The use of digester gas in a heat pump for hot water production is another possibility, but so far as is known, not tested. Digester gas can be used in suitably modified gas fires for space heating, or in gas rings or gas cookers or refrigerators, or in any other way that natural gas is used, for instance in ducted hot-air heating systems.

A gas boiler can supply hot water for factory use, space heating in houses, factory or animal houses, and on farms for washing milking equipment and other objects. Another use for digester gas on farms, particularly in the summer and early autumn seasons, when demands for space heating and digester heating are minimal, is for crop drying. Crop driers use oil, gas, LPG, and even coal, burners to produce hot air which is blown over the crop as it passes through a tunnel kiln. A dual-fuel burner system could allow digester gas to be used for at least part of the fuel needs. Such systems are projected. A large digester system would be needed for crop drying, though, as energy inputs are high. In the drying of one tonne of grain approximately 299 kWh of heat are needed to reduce the moisture content by 10 %.

This section has dealt briefly with some uses of digester gas, but one that has not been mentioned is use in vehicles. While, as has been shown, digester gas is a good fuel for compression or spark-ignition engines, it was shown that large volumes of gas at low pressure are needed. While this can be obtained from a gas-holder connected to a continuously running digester, a vehicle engine must carry the gas with it. The volume of gas can only be accommodated in high-pressure cylinders. The weight penalty of such cylinders is enormous. The cylinders required to carry digester gas equivalent to the 10-gallon petrol tank of a small car would be a large fraction of the pay-loading of the car. High-pressure digester-gas cylinders were used on some tractors driven by the early German digesters previously mentioned, and they were used on lorries working in and around sewage works at one time. But in these cases journeys were short, so little fuel was required, the vehicles were large, and in the case of tractors weights may normally be added to improve wheel adhesion. High pressure cylinders on cars would have to be changed when refuelling as provision of cylinder-charging facilities could not be available at every fuel stop. The routine testing of high-pressure cylinders is also a consideration. Nevertheless, as described elsewhere, digester gas is planned as fuel for cars, the disadvantages being set against those of a country (New Zealand) with no indigenous oil supplies.

SAFETY PRECAUTIONS AND TESTS WITH DIGESTERS

Methane is an odourless gas, and 5–15 % methane in air forms an explosive mixture. Methane itself is lighter than air, but the digester gas mixture with carbon dioxide is denser than air and the two gases do not easily separate. Although not poisonous, methane is anaesthetic in a concentration of 1000 ppm in air. Digester gas, like natural gas, can then be a source of danger, but in practice the dangers can be overcome with suitable precautions.

The very largest digester plants designed to produce purified methane for state gas pipelines or electricity on a large scale, will be run as 'factories', and should have all necessary safety devices on digesters, compressors and other plant. They will also have trained workers running them and be under the control of resident engineers. The following notes are mainly concerned with smaller plants.

Although methane is odourless, as is carbon dioxide (except in very high concentration when it 'tastes' slightly acid), digester gas contains small amounts of impurities which give it a smell. So its presence will usually be detected. The presence of small amounts of methane (and other gases) in air can be monitored by some of the commercial sets on the market in which a hand-operated bellows sucks a fixed volume of air into a test apparatus to give a visual read-out in a change in a graduated tube, for instance. More sophisticated meter-readout detectors for methane are available, but these are expensive. For small digester plants the smell of the gas will be sufficient to indicate its presence, but the method of running should be such as to avoid possibilities of danger. There is, however, one problem about smell, and that is that since the impurities and the methane and carbon dioxide in digester gas are also generated in undigested animal faecal slurries that have been standing, the slight odour around slurry tanks could be mistaken for, or mask, the smell of digester gas. However, as shown by occasional accidents from gaseous discharge when opening tanks of slurry that has been stored for some time, especially in confined spaces, the head-space in any slurry tank, or the air immediately over an open tank, should be treated as potentially as dangerous as digester gas. (These dangers have been explained in books on animal waste management; e.g. Hobson & Robertson, 1977).

The size of digesters almost entirely precludes them from being placed inside a building, so that the normal winds should rapidly dilute and disperse any gas leaking from the digester or gas-holder. As previously explained, a water-sealed gas-holder, a floating digester top or a weir or standpipe overflow, will act as a safety valve for overpressure in a digester. Gas blowing off in this way should be rapidly dispersed in the air, and will be detectable by bubbling sounds from the digester or gas-holder, or sludge frothing out of a weir. Of course, danger exists close to the gas escape and no smoking must take place nearby, but smoking should generally be discouraged around a digester plant.

Fixed-top digesters, with pumped or other closed outlet, require some form of safety valve. The difficulties of using a spring-loaded valve in a corrosive, wet atmosphere and where particles of digesting sludge may clog it, have already been mentioned. Some types of commercial safety valve are fitted to sewage digesters, but the bursting disc may be a solution. If the digester is not being run under much pressure, then the safety valve can consist of a large-bore water or oil manometer which can blow-out with overpressure.

Particularly on the larger plant, excess gas produced is preferably flared off. Gas flares should be set up as many yards as possible from the digester plant and are usually automatically ignited. The ignition signal can come from a level switch at the highest rise of the gas-holder, or from elsewhere (e.g. a pressure switch on a pressurised gas-holder). Commercial gas flares for small gas flows are available. The flare has, of course, to ignite and burn in the open in all weather conditions.

Engine and boiler houses should be well ventilated to prevent any leaking gas building up in corners. And, as is usual in any boiler or engine house, exhaust gases should be led outside through properly designed and installed pipes and flues.

Such precautions should obviate any chance of explosion or fire and also chance of poisoning from hydrogen sulphide in the gas. Hydrogen sulphide is extremely poisonous and a concentration of less than 3 ppm is the maximum allowed for continuous exposure of personnel. A concentration of 20 ppm produces toxic symptoms on quite short exposure. Even if it does not cause unconsciousness, a low concentration of hydrogen sulphide can cause nausea, headache and dizziness. Although precautions other than the good ventilation will not be needed for engine houses, etc., in a digester plant, it is possible that if digester gas became generally used in houses, legislation might insist that the hydrogen sulphide content of the gas be reduced to that allowed for domestic gas. This in Britain is 3·3 ppm. To get digester gas to this level would mean scrubbing the gas as previously described. However, since this would make the gas virtually odourless a 'smelly' compound might have to be added to aid detection of the gas, as is done with domestic natural-gas supplies.

A further precaution against explosions is the provision of flame traps in gas pipes, between a boiler, engine or other equipment and the gas-holder. A flame trap will prevent an explosion flame from travelling back down the pipe. Traps are commercially available and consist of a flame detector, a flame retarder (some form of gauze disc) and a fail-safe magnetic valve in the pipe line. The flame detector actuates the cut-off valve before the retarded flame can pass the valve.

The usual precautions about handrails to digester tops, walk-ways and ladders on digesters and suitable screening for moving shafts of engines and pumps should be observed. In addition, digester sites on farms should be fenced to keep out children, animals and casual passers-by. Lagoons or open tanks storing digested sludge or feedstock should also be fenced. Operation of tractors fitted with rakes or scrapers might also be discouraged near digesters in case a sudden turn smashes a rake against the digester.

If a digester has to be emptied, precautions should be taken to avoid build-up of gas from the discharging sludge. Once the digester is emptied all possible man-holes should be opened to let natural ventilation blow away gas inside the tank. This should be continued for some days, particularly if there is a residue of grit or other sludge solids in the bottom of the tank. If man-holes are small compared with the tank volume, then forced-air ventilation should be used, Preferably some test of the gases in the tank should be made before anyone enters it, and safety lines, etc., may in any case be required unless there is a large man-hole near the bottom through which workers can get in and out easily and quickly. Apart from any danger of asphyxiation of workers, residual gas could form an explosive hazard, so the greatest possible care in ventilation should be used before any cutting or welding is done on a digester tank. And solid residues in tanks always form a possible source of continued gas generation and should be removed as the first operation on an emptied digester.

While the previous section may have given the impression that a digester is a potentially dangerous piece of equipment, in actual practice running a digester presents no hazards provided a few obvious precautions are taken. The Aberdeen digesters have been running for seven or eight years on a farm site without incident. The main precautions need to be taken when some alteration to the design of digester and fittings is to be made; or the digester is to be emptied, for instance.

A commercially made digester will be started up and brought to stable operating conditions by the manufacturers. Once stability is obtained there will be, as pointed out elsewhere in this book, little possibility of microbiological breakdown, even if there are periods of no loading, or under- or (within very broad limits) over-loading. Similarly a long period without heating will not cause permanent breakdown, but a few hours at a temperature of 45 °C or over could kill the bacteria in a mesophilic digester, just as 65 °C is probably about the temperature limit for a thermophilic digester. However, for a consistent output of gas a digester needs to be run in a consistent manner. The biggest digester plants will, as has already been indicated, be run by specialist staff and will have test-equipment, workshop facilities, and so on, commensurate with what they are—factories producing biogas. The following notes are meant for the smaller digester plant, say a farm unit, where staffing can only be part-time and no specialist personnel or test equipment are available.

The automated digester we have been considering will have been built for the particular site and its feedstock production rate. The time-on of input pumps will, therefore, be set by time-clocks and only an occasional check on these and a check on running of pump motors will be necessary. Similarly, timings of digester and input tank stirring will have been set. Overhaul of pumps and motors, lubrication, etc., will be infrequent and either by makers' recommendations or general observation. Unless there is a reason why gas output must be absolutely continuous, duplication of input and other pumps is hardly worthwhile. The input tanks will, as previously mentioned, act as ballast tanks for a day or two's feedstock and the digester will not suffer if not loaded for a day or two. Although, if the digester is running at its correct detention time, gas production will fall off in a day or so if there is no loading.

Gas-production rate is the easiest measurement to make on a digester and the one which gives much information. Gas production rate will vary a little from day to day, but should, on average, be constant. If production declines steadily over some days, then a reason should be looked for. Apart from a leak somewhere, the feedstock input rate should be looked at and also the feedstock consistency. On a farm site a leak of water into slurry channels may have occurred, lowering the solids content of the waste. Such a reduction in solids may be obvious, but simple hand-centrifuge or slurry settling tests are possible, which can be done either as routine or in case of change in gas production and will give a more exact measure of solids content in the feedstock. If feedstock input and composition have not varied, other causes will have to be looked for. Digester temperature and temperature of water in heater pipes should be monitored, either automatically or by manual reading once a day or so. A sudden input of detergent or disinfectant may have accidentally occurred. If such a cause can be found and its input stopped, or if it has stopped, then continued running of the digester should dilute out the 'poison' and the digester will gradually get back to normal running. But, of course, if a large volume of feedstock has been contaminated this will have to be disposed of in some way and replaced by uncontaminated feedstock. If input of a 'poisonous' (to bacteria) substance is suspected then continued running on uncontaminated input is the best advice; stopping the digester feed will only allow the 'poison' more opportunity to destroy the bacteria.

If poisoning is not a reasonable cause then stopping loading for two or three days followed by gradual increase to normal loading over a few days will in most cases allow a digester to self-correct the effects of a previous overloading or some other cause of upset.

A sudden increase in gas production (probably accompanied by engine or boiler malfunction) will most likely be due to air getting into the system through an input pump or pipe, perhaps. If not allowed to continue this will do little harm to the digestion and when the leak has been stopped loading can continue as usual.

Gas production can easily be measured, but gas composition cannot without special equipment. An indication of carbon dioxide content can be obtained by shaking a known volume of the gas with a little dilute caustic soda solution, but this will not tell if the residual gas is methane, nitrogen or air. However, in general the composition of the gas will be unchanged even if the production rate changes considerably (except in the case of air intake). Production rate is a good guide to digester function and it is hardly worthwhile for the smaller digester owner to buy gas analysis apparatus.

The concentrations of volatile fatty acids, ammonia, BOD, COD, etc., mentioned elsewhere cannot be measured except with laboratory apparatus. The pH of samples of digester overflow could be measured with a portable meter, but such meters are not cheap, and unless the overflow sample is tested quickly and re-equilibrated under a carbon dioxide atmosphere measurements will be accurate enough to detect only a large pH change. Electrodes for pH measurement fixed into a digester suffer from problems of fouling and are not reliable. An approximate indication of pH of the overflow would be given by some of the indicator papers available. But in the authors' experience pH changes in farm-waste digesters are rare and, again, gas production would give just as good or better an indication of digester malfunction as pH change.

The running of an engine or boiler will be governed mechanically and such apparatus will have cut-outs to prevent damage or danger from apparatus or digester malfunction (e.g. gas cut-off in the event of pilot-light failure on a boiler). Operation of various safety devices or indicators of digester overheating, a filled gas-holder, and so on, can be made to control alarms if required, or, only some malfunctions might operate alarms. For instance overnight cooling of a digester due to boiler or valve failure is undesirable but would cause no permanent damage to the digestion, but overheating for 10 hours at night could kill the bacteria and a complete start-up procedure with loss of gas production for some weeks would be the necessary result. So an alarm to indicate a malfunction of boiler thermostat or water circulation and bypass system would be better to give a 'high' alarm than a 'low' alarm if only one were to be provided.

For the larger digester, microprocessor control could be considered, with printout of temperatures and other parameters and either activation of alarm signals or shut-down of engine or other functions in the event of some malfunction.

Finally, to return to the question of gas hazards; wind dispersion makes it very unlikely that explosive amounts of methane or toxic amounts of hydrogen sulphide would be found in the open around a digester. The proper ventilation already referred to should prevent accumulations inside buildings. The smell of hydrogen sulphide in the gas, particularly on first entering a building, should be sufficient warning of possible dangers. Hydrogen sulphide can be tested for in very small concentration by simple tests such as discolouration of damp lead acetate paper, but unless the exposure of the paper is controlled and calibration papers are available, determination of the concentration of gas is impossible. Commercial detectors giving a visual reading of concentration of hydrogen sulphide by change in an indicator tube are available. Purchase of one of these might be considered worthwhile as it could also indicate the presence of methane in that if concentration of hydrogen sulphide is high then it is a reasonable deduction that the methane constituent of the digester gas is also present. Comparison of the concentration of sulphide in the air sample with that of the concentration in the digester gas itself will give the required information. The concentration of sulphide in the digester gas will, like that of methane, remain relatively constant if the digester feedstock is of constant composition.

The running of a farm or similar digester should not be a time-consuming task, perhaps one-half man hour per day, and if a commercially-made digester is installed the builders will be able to give advice on the necessary safety devices and what to do in case of malfunction. More experience on digester running is being obtained every day.

REFERENCES

A Chinese Biogas Manual (1979). Translated by M. Crook. Intermediate Technol. Publishers, London.

A Symposium: The Starting-up of Digesters (1964). *J. Inst. Sew. Purif.*, 303.

ANON. (1951). Defu-Mitteilungen, Heft 9.

BADGER, D. M., BOGUE, M. J. & STEWART, D. J. (1979). *N.Z. J. Sci.* **22**, 11.

BRADE, C. E. & NOONE, G. P. (1979). Paper presented at IWPC AGM, University of Aston.

BURGESS, S. G., GREEN, A. F. & WOOD, L. B. (1957). *J. Inst. Sew. Purif.*, 206.

CHESSHIRE, M. (1978). In: *Proc. Seminar Anaerobic Digestion of Farm Wastes*, ADAS, Reading.

CHUNG PO (1973). *Proc. Int. Biomass Energy Conf.*, Biomass Energy Inst. Inc., Winnipeg, Canada.

DODSON, C. E. (1978). In: *Proc. Seminar Anaerobic Digestion of Farm Wastes*, ADAS, Reading.

DOHNE, E. (1980). In: *Anaerobic Digestion* (eds. D. A. Stafford, B. I. Wheatley and D. E. Hughes) Proc. 1st Int. Symp. An. Dig., Cardiff. Applied Science Publishers, London.

DREW, E. A. & SWANWICK, J. D. (1962). *Publ. Wks. Mun. Serv. Cong.*, p. 1.

ECOTOPE GROUP (1977). Dynatech Rept. 1683, Dynatech Co., Camb., Mass.

ECOTOPE GROUP (1979). Dynatech Rept. 1883, Dynatech Co., Camb., Mass.

ESCA (1976). Tech. Note 141C/A East of Scotland Coll. Ag., Edinburgh.

FINCH, G. (1956). *Water Sew. Wks.*, **103**(6), 276.

FRY, L. J. (1974). *Practical Building of Methane Power Plants*. D. A. Knox, Andover, England.

FULLEN, W. J. (1953). *Sew. Ind. Wastes* **25**, 576.

GARBER, W. F., O'HARA, G. T., COLBAUGH, J. E. & RAKSIT, S. K. (1975). *J. Water Pollution Control Federation*, **47**, 950.

Gobar Gas—Why and How (1975). Director of Gobar Gas Scheme Pub., Khadi and Village Industries Commission, 'Gramodaya', Irla Road, Vile Parle (West), Bombay 400 056.

HAGA, K., TANAKA, H. & HIGAKI, S. (1979). *Ag. Wastes* **1**, 45.

HAYES, T. D., JEWELL, W. D., DELL'ORTO, S., FANFONI, A. P., LEUSCHNER, A. P. & SHERMAN, D. F. (1980). In: *Anaerobic Digestion* (eds. D. A. Stafford, B. I. Wheatley and D. E. Hughes) Proc. 1st Int. Symp. An. Dig., Cardiff. Applied Science Publishers, London.

HEMENS, J. & SHURBEN, D. G. (1959). *Food Trade Rev.* **29**, 2.

HILLS, D. J. & STEPHENS, J. R. (1980). *Ag. Wastes*, **2**, 103.

HOBSON, P. N. (1979). In: *Straw Decay and its Effect on Disposal and Utilisation* (ed. E. Grossbard). Wiley, London, p. 217.

HOBSON, P. N., BOUSFIELD, S. & SUMMERS, R. (1974). *Critical Reviews in Environmental Control* **4**, 131.

HOBSON, P. N., BOUSFIELD, S. & SUMMERS, R. (1979). *Proc. Ann. Res. Meet. Inst. Chem. Eng.*

HOBSON, P. N. & MCDONALD, I. (1980). *J. Chem. Tech., Biotech*, **30**, 405.

HOBSON, P. N. & ROBERTSON, A. M. (1977). *Waste Treatment in Agriculture.* Applied Science Publishers, London.

HOBSON, P. N. & SHAW, B. G. (1973). *Water Res.* **7**, 437.

HOBSON, P. N. & SHAW, B. G. (1976). *Water Res.* **10**, 849.

HOWARD, B. (1979). *Ag. Wastes* **1**, 11.

IRVING, S., KING, R. & BULL, D. (1978). *Chem. Eng.* (Nov.) p. 831.

JEWELL, W. J. (1980). In: *Anaerobic Digestion* (eds. D. A. Stafford, B. I. Wheatley and D. E. Hughes) Proc. 1st Int. Symp. An. Dig., Cardiff. Applied Science Publishers, London.

LETTINGA, G., VAN VELSEN, A. F. M., DE ZEEUW, W. & HOBMA, S. W. (1980). In: *Anaerobic Digestion* (eds. D. A. Stafford, B. I. Wheatley and D. E. Hughes) Proc. 1st Int. Symp. An. Dig., Cardiff. Applied Science Publishers, London.

MECKERT, J. W. (1978). *Inst. Gas. Technol.*, August.

METANGAS UR GÖDSEL (1976). Jordbrukstekniska Institutet, 75007 Uppsala 7, Sweden.

MILLS, P. J. (1977). In: *Proc. 9th Ann. Waste Manag. Conf.*, Cornell.

MILLS, P. J. (1979). *Ag. Wastes* **1**, 57.

MILLS, P. J. & MONTGOMERY, P. (1979). Private Report.

MORRIS, J. E. (1980). In: *Anaerobic Digestion* (eds. D. A. Stafford, B. I. Wheatley and D. E. Hughes) Proc. 1st Int. Symp. An. Dig., Cardiff. Applied Science Publishers, London.

MOSEY, F. E. (1980). In: *Anaerobic Digestion* (eds. D. A. Stafford, B. I. Wheatley and D. E. Hughes) Proc. 1st Int. Symp. An. Dig., Cardiff. Applied Science Publishers, London.

NELSON, E. D. & BAILEY, J. F. (1979). *Water Sew. Wks* (Feb) p. 30.

NEWELL, P. J., COLLERAN, E. & DUNICAN, L. K. (1979). Poster presentation at 1st Int. Symp. An. Dig., Cardiff.

NEYELOFF, S. & GUNKEL, W. W. (1975). In: *Energy, Agriculture and Waste Management*, Ann Arbor Sci. Publishers, Michigan.

ONAN Tech. Bull. (1974) T-015.

PATEL, J. J. (1951). *Poona Agri. Coll. Mag.* **42**, 150.

PATEL, J. J. (1959). Paper presented at Indian Agric. Fair. Symp.

PATEL, J. J. (1963). *Gobar Gas Plants.* Khadi Gram Udyog, India.

PATEL, J. J. (1964). *Installation of Biogas Plant.* Bombay Khadi Village Industrial Commission.

PFEFFER, J. (1979). Dynatech Rept. 1845, Dynatech Corp., Camb. Mass., p. 83.

PICKEN, D .J. (1978). In: *Proc. Seminar on Anaerobic Digestion of Farm Wastes,* ADAS, Reading.

PICKEN, D. J. (1979). Unpublished communication at a seminar in Aberdeen. Rowett Research Institute, Aberdeen.

PRETORIUS, W. A. (1971). *Water Res.* **5**, 681.

ROBERTSON, A. M., BURNETT, G., BOUSFIELD, S., HOBSON, P. N. & SUMMERS, R. (1975). In: *Proc. 3rd Int. Symp. Livestock Wastes,* Univ. of Illinois, USA. p. 544.

Safety in Sewers and Sewage Works (1969). Inst. C. E. Min. Health and Local Govt.

SAMBIDGE, N. (1972). *Water Pollution Control Federation* **44**, 105.

SATHIANATHAN, M. A. (1975). *Biogas, Achievements and Challenges.* Assoc. Volunt. Agencies Rural Dev. A/1 Kailash Colony, New Delhi-110048.

SCHROEPFER, G. J., FULLEN, W. J., JOHNSON, A. S., ZIEMKE, N. R. & ANDERSON, J. J. (1955). *Sew. Ind. Wastes* **27**, 4.

SHAW, B. G. (1971). Ph.D. Thesis, University of Aberdeen.

SOLIMAN, H. A. (1978). Ph.D. Thesis, University of Leicester.

STEFFEN, A. J. (1953). *Biol. Treat. Sew. Ind. Wastes* **11**, 126.

STEWART, D. J. (1980). In: *Anaerobic Digestion* (eds. D. A. Stafford, B. I. Wheatley and D. E. Hughes) Proc. 1st Int. Symp. An. Dig., Cardiff. Applied Science Publishers, London.

SU-TE SHIAN, MIN-CHING CHANG, YUAN-TAW YE & WEI CHANG (1979). *Ag. Wastes* **1**, 247.

SUMMERS, R. & BOUSFIELD, S. (1976). *Proc. Biochem.* **11**, 3.

SWANWICK, J. D., SHURBEN, D. G. & JACKSON, S. (1969). *Water Poll. Control* **6**, 639.

TAYLOR, D. W. (1972). In: *Proc. 3rd Nat. Symp. Food Proc. Wastes.,* New Orleans, USA.

TENNEY, M. W. & BUDZIN, G. J. (1972). *Water Wastes Eng.* **9**, 57.

VERHOFF, F. H., TENNEY, M. W. & ECHELBERGER, W. F. (1974). *Biotech. Bioeng.* **16**, 757.

WHEATLEY, B. (1978). In: *Proc. Seminar Anaerobic Digestion on Farm Wastes,* ADAS, Reading.

WIEDEMANN, F. (1977). *Gas-U-Wasserfach (Wasser, Abwasser)* **118**, 278.

CHAPTER 6

Uses of Digested Sludge

Whether the primary function of a digester is reduction of pollution or production of energy the process will produce a residue in the digested feedstock. Since, except for the 'solid' digestions previously mentioned, the feedstock is very largely water, decrease in volume due to gasification of solids will be small and the digested sludge will be of much the same volume as the feed. Digested sludge from some sewage works is just dumped on land or at sea and so a transport cost is involved in removing it. This is part of the overall cost of sewage purification for the city and is covered by city rates. But for the farmer or factory using a digester for pollution control and production of energy, dumping of digested sludge adds to the process an unwelcome cost. The same argument applies to digesters run primarily for large-scale gas production. If the sludge can replace something bought in to the farm or factory, or if it can be sold off-site, then this provides a monetary return which can decrease the cost of the pollution control or gas production. Finding a use for digested sludge is thus an integral part of the production of energy from wastes and it is also relevant to present-day thinking on recycling and maximum use of resources.

It can be argued, with reason, that if the digester feedstock is a waste that has in any case to be got rid of by dumping, then the cost of dumping the digested waste can be ignored. If gas can be obtained before the waste is dumped then only the digester plant should be considered in costing the gas. Pfeffer (1973) used this argument in a model of digestion of city garbage and sewage sludge. The digestion gave a product which could form better landfills than the original garbage, which gave a monetary benefit.

Some farms have sufficient land for spreading their animal excreta as fertiliser. The fertiliser value of digested excreta is the same as that of the raw excreta, so there is no monetary gain here. The value of the digestion is again that gas is obtained between production and disposal of the waste. But, as with the garbage there is a gain in that digested sludge is easier to pipeline or tractor spread than the original. And in the case of farms close to other habitation or to water supplies the digested excreta is overall less polluting and much less liable to cause complaints about odours. In addition it can be stored with less inconvenience.

But whether or not the cost of disposal of the digested sludge can be considered in this way there are many reasons for seeking uses for the sludge which will give positive returns, and some of the uses will be considered in this chapter.

USE AS FERTILISER

The most common use of digested sludge is as fertiliser. A primary benefit of the small Gobar- or Chinese-type digester is that they produce a fertiliser as well as energy, unlike the burning of cow-dung, and this fertiliser contains less pathogenic organisms and weed seeds than untreated animal excreta.

In the case of digested sludge from the sewage plants of large cities the volumes have been so great and the demand for fertiliser so small that in many cases only a fraction of the sludge has been used as fertiliser and the rest has been dumped. However, the increasing costs of chemical fertilisers, in both monetary and energy terms, and the need for conservation of resources is leading to a reversal of this thinking on disposal of sludge and much more is now being used as fertiliser. Indeed, at one time bulk sludge was given away as fertiliser to anyone who would take it, or sold at transport price, but it can now be sold on a more realistic basis.

Digested sludge in bulk can be delivered wet to farmers and market gardeners and spread by pipeline from central tanks loaded at intervals by the sewage works' tankers, or it can be spread by the road tankers themselves or transferred to tractor-drawn tankers. For use by household or small market gardeners a dry, easily-handled material is needed. Sewage-works sludge has conventionally been dried by spreading on to drying beds of clinker and sand or other porous fill, where it can drain and dry in the wind for many weeks. In the authors' experiments digested piggery slurry has been shown to dry to a blackish, odourless, friable mass on small-scale, experimental, drying beds. The water from drying-bed drains is, of course, still polluted. Such drying is subject to climatic vagaries, but is less costly in both money and energy than the various chemical or physical coagulation and settling methods, followed by filter or centrifuge dewatering and final drying, that can be used. Dried sludge can be bagged and sold in small amounts to domestic gardeners and others at a much higher price than bulk sludge so that there is a better return, although the amount that can be disposed of in this way is rather limited. Dried sludge, because 90 % of its weight has been removed, is cheaper on an NPK basis to distribute by road than is wet sludge.

Because of its ease of handling and storage and because it does not add to the water content of already wet ground and so lead to polluting run-off to rivers or underground water reservoirs, dried sludge has advantages over the wet material. In the case of both farm and municipal sludges more thought is being given to the production of a dry material. But the big problem remains; the sludge is 90% or more water and removal of this water, unless done on drying beds, is costly and energy-consuming. The sludge can be 'thickened' by standing, when the supernatant water can be drawn off from the top of the tanks and a sludge of higher solids content removed from the bottom of the tank. This is the usual practice in sewage works. Water from the top of the sludge is still polluted and contains fine particles, so it must be returned for further purification before being run to a river. Such settling requires some 10 or 20 days, and a more rapid and complete settlement can be obtained by the use of coagulating agents. These are tannins, synthetic polyelectrolytes, ferric sulphate, ferric chloride, aluminium chloride, aluminium chlorohydrate, ferrous sulphate ('copperas') and chlorinated copperas. The best type of reagent to use should be determined for the particular sludge involved. The drawback to the use of coagulating agents is that they add to the cost of the process, and other processes of electrolytic or air flotation coagulation can be even more expensive.

Even the chemically-coagulated sludge is still of high water content and so presents problems in further dewatering. If this is not done on drying beds, mechanical methods can be used. The sludge can be brought to a reasonably dry condition by filtration on vacuum drum filters, or filter presses, or band filters. But all these, like centrifugal dewatering, are expensive in capital and running costs. A cheaper method of dewatering digested sludge might be found in the use of the solids-separating machines previously mentioned as used on farms for separation of solids from animal slurries. However, the main fertiliser value of the sludge lies in the bacteria and such separators would have to be adapted to take out much more of the fine particulate matter than they do at present. Experiments on these lines are just beginning.

Sludge dewatered by mechanical means, although a solid, will still have a relatively high water content (up to 70% or more). While properly-digested sludge solids should have little or no odour and should not be subject to further putrefaction, they could be substrates for mould growth, and so the material would have to be distributed and used relatively quickly and it would be unsuitable for bagging and selling in small quantities. Further drying to about 15% moisture content or less (at which mould growth is inhibited) would require breaking up of the material and hot air or other drying.

What can be done with the digested sludge depends ultimately on the value of the product as fertiliser and on the use to which the gas produced in the digestion is being put. For instance, if a farm digester were producing more energy than was required, unused electrical power could be used to drive a solids separator to dewater the sludge and excess engine heat or gas could be used to dry the dewatered material. A commercial piggery digester is being used to power a plant which gives a dried-sludge fertiliser and a highly purified water. The local regulations on pollution require the water discharged to rivers to be purified and the dry fertiliser can be sold because it does not cause the run-off pollution problems associated with liquid sludge.

As previously pointed out, digested sludge or the supernatant water from settled sludge is not of high enough standard in BOD or other parameters to be normally discharged to a water course, but the processes of concentrating the solids for use as fertiliser can bring about various degrees of cleaning-up of the water so that it can be used, if not directly discharged. Some examples from the Aberdeen work on piggery slurry will be used to illustrate this.

Digestion removes some of the cellulose, hemicellulose, starch, pectins and other carbohydrates, and fats, from slurries of animal excreta. The digested sludge contains lignified fibres undegradable anaerobically and only slowly degraded by aerobic micro-organisms. This fibrous material has of itself no, or little, value in terms of fertiliser NPK, but acts as a source of humus and as a soil conditioner. The NPK value of the sludge resides mainly in the finer particulate matter, the bacteria, although, of course, in practice some of the bacteria will be adherent to the fibrous material. The nitrogen is in the form of ammonia, much of which appears to be in, or absorbed to, the bacterial cells, and in the form of protein and other constituents of the cells. It will thus be available at different rates to plants in the same way as is the nitrogen in untreated excreta. The phosphorus and potassium are again mainly associated with the bacterial cells. The sludge may be valued as a fertiliser on its nitrogen content, but in some cases phosphorus may be the more important plant nutrient.

A typical analysis of the dry solids obtained by lightly centrifuging digested piggery waste is given in Table 6.1 (Summers & Bousfield, 1980). This light centrifuging was the equivalent of settling for some days and gave a thickened sludge of about 7–7·5 % TS (i.e. about two to three times the original).

The supernatant liquid from the centrifuging was still blackish in colour

TABLE 6.1

AN ANALYSIS OF DIGESTED
PIGGERY WASTE SOLIDS
(from figures given by
Summers & Bousfield,
1980)

Lipid	9·42%
Cellulose	13·25
Hemicellulose	16·26
Lignin	11·75
Crude protein[a]	25·75
Ammonia N	2·68
P	2·61
K	1·17
Ash	15·43
Starch	0·00

[a] Non-ammonia N × 6·25.

and contained ($\%$ w/v): Total N 0·17; NH_3N 0·14; P 0·006; K 0·08 (about 0·014 as P_2O_5 and 0·1 as K_2O). The analysis of the solids would be much the same with any waste from pigs housed on slatted floors with no bedding (or only a little sawdust) and fed on a dry grain diet. The analysis of the whole digested slurry would then depend on its solids content. For example, if the slurry entering the digester had a Total Solids content of 7 %, then this would be reduced by about 40 % on digestion (i.e. a reduction in organic matter of about 57 % as the TS includes some 30 % salts and grit). The TS of the digested sludge would be 4·2 % or 42 kg/m³ and this would contain 2·8 kg N, 2·5 kg P (as P_2O_5) and 0·59 kg K (as K_2O). The rate of application and fertiliser value could be calculated from these figures.

The analysis of the digested sludge also shows that somewhat over 40 % is fibre. Comparison of various analyses of raw and digested pig waste showed that some 41 % of the cellulose and 48 % of the hemicellulose was gasified and the final fibrous material had a higher lignin content than the original. The other main constituent removed in the digestion is fat, and the previous calculations also showed that some 53 % of the fat was digested. Much of the remaining fat would be the lipid material of the bacteria.

The thickened sludge used in the analyses of Table 6.1 was freeze-dried, so that some of the ash would come from salts in solution. If water were removed from the sludge by filtration or high-speed centrifugation these salts would not be included in the dried material. If the bulk of the fibrous material could be removed by mechanical separation, then the remaining suspended solids would be largely bacteria and the figures in Table 3.1 show that this would have a nitrogen content approaching 10%, which is the generally accepted figure for pure dried bacteria.

These figures, then, give an idea of the value of digested pig slurry as a fertiliser and nitrogenous feed additive. In fertiliser terms the proportions of N, P and K would class the solids as a high-nitrogen, high-phosphate fertiliser. A general-purpose fertiliser has N, P_2O_5 and K_2O in proportions of 1:1:1, a high-nitrogen in proportions 2:1:1, and so on. Reliable average figures for many samples of digested cattle and poultry wastes have not yet been obtained by the authors. Figures quoted for the analysis of domestic sewage sludge have varied quite widely, possibly because of varying contents of aerobic sludge and of the material added to the sewers of the different towns from which the sludges came (a typical summary is given by Hobson *et al.*, 1974). Methods of analysis have also varied. The composition of the digested sludge will also be variable.

The results of various secondary treatments of digested pig slurry (Table 6.2) show how some methods might be used to concentrate the sludge and improve the quality of the supernatant water (Summers & Bousfield, 1976, 1980). Further experiments might improve the results of aeration, but the analysis of the final liquid depends to some extent on the input. A similar aeration treatment of a more dilute digested sludge gave a supernatant liquid with a BOD of 50 (Hobson & Shaw, 1973). Aeration seems probably the best treatment for reducing the BOD of the digested sludge, but mechanical aeration is costly both in monetary and energy terms. Unless ammonia was stripped off by the aeration or removed by denitrification then the nitrate content of aerated digested sludge could also be a problem for disposal of the treated water. A commercial piggery-waste digester system previously mentioned uses chemical coagulation, electrolytic flotation and ozone treatment of the water to effect a high degree of purification.

It is common practice to add copper (as copper sulphate) to pig feeds as a growth promoter. This means, as shown previously, that the digested sludge contains copper sulphide. The copper, at a concentration usually of 200 ppm in the feed becomes concentrated in the sludge solids. Some analyses (Hobson & Shaw, 1973) showed 850 ppm of copper in dry, digested-piggery-sludge solids. This copper would, of course, be reduced in concentration if digested sludge were compounded into a feedstuff, but might make such digested pig sludge suitable only for addition to pig feeds. Since the copper sulphide is insoluble it has been suggested that it would not be taken up by plants if digested pig slurry were used as fertiliser and that

TABLE 6.2
VARIOUS SECONDARY TREATMENTS OF DIGESTED PIGGERY WASTE
(from Summers & Bousfield, 1980)

	BOD (mg/litre)	% reduction	COD (mg/litre)	% reduction	VFA (mg/litre)	NH$_3$N (mg/litre)	TS (%)
Digester							
Input	17 543	0	66 031	0	3 697	1 908	3·4
Output (whole)	2 787	84	34 133	48	490	1 781	2·1
Secondary treatments[a]							
Settling only (14 days)[b]	650	96	34 750	47	432	1 480	1·0
Flocculation with Polyelectrolyte[c]	387	98	1 075	98	125	690	0·6
Aeration (for five days)[b]	210	99	814	99	420	84	1·0
Some settlement and coke-bed filtration[b]	1 480	92	9 142	86	424	1 686	0·9

[a] The analyses are for the liquid fractions.
[b] Supernatant liquid dark brown–black.
[c] Supernatant liquid water-clear and only faint brownish colour.

the particles would gradually settle into the soil. Unless heavily-fertilised grass were immediately grazed by cattle, when there might be chance of them ingesting dried sludge, there would seem to be no danger from using digested waste from copper-fed pigs as fertiliser. (See for instance experiments by Gracey *et al.* (1976) using raw slurry.) A couple of weeks after the sludge has been spread should be sufficient to wash off sludge from grass, in the usual British weather conditions.

Experiments by the College of Agriculture in Aberdeen (briefly mentioned by Summers *et al.*, 1980 and Hobson *et al.*, 1980) showed that spreading of digested piggery slurry by tractor-tanker did less damage to land than did spreading raw slurry, particularly at high application rates. For instance, at an application rate of $44\,m^3/ha$ the crop (grass) yield from the wheel tracks of the spreader was 55 % down on that of the rest of field when raw slurry was spread, but only 12·5 % down when digested sludge was spread.

When animal or human excreta are used as fertiliser, questions of possible spread of pathogenic organisms arise. Experiments tend to show that anaerobic digestion is generally destructive to pathogenic organisms, but a full range of tests with different organisms and different digestions has not yet been carried out. Ward *et al.* (1976) found that municipal digested sludge inactivated poliovirus at a rate proportional to time and temperature. The virucidal activity against *Reovirus* was found to be greater in digested sludge than in raw sludge (Ward & Ashley, 1977). These are only examples and the question of destruction or inactivation of pathogens in a material such as sludge is complex and different results may be found under different conditions.

The authors have carried out some tests with salmonella and piggery-waste digesters, as the spreading of enteric disease amongst humans through the agency of contact with animals or, more generally, animal products is now regarded as increasingly serious, and this obviously involves spread of infection amongst animals.

Previous tests (e.g. Shaw, 1971) had shown that salmonella were not found in the digester effluents, but that they were also virtually absent from the piggery wastes used over the years. So other tests were done with a digester loaded with salmonella. A strain of *Salmonella anatum* was used.

It was shown that digestion considerably reduced the number of salmonella, even when these were introduced into a digester in numbers which would be attained in practice only in the case of a large outbreak of disease in a pig herd (Summers & Bousfield, 1978; Hobson *et al.*, 1980).

Salmonella were added to a pilot-plant digester running at 35 °C and a detention time of 10 days to give a digester population of 1.7×10^5/ml. After three days their number had decreased by 98 %. The half-life of the bacteria was only a few hours and death was exponential. Continued monitoring showed that although numbers in the digester were small, the bacterial population remained higher than a continued exponential death would suggest. It was shown that this could have been due to reinoculation of the digester contents from growth of bacteria on a slurry crust on the wall of the digester above the surface of the contents. This digester was stirred by a rotating paddle and surface turbulence was small. The possibilities of wall-growth such as this would probably be considerably lessened by gas mixing which causes surface turbulence. The number of salmonella escaping in a digester overflow would also be considerably reduced if the sludge were stored before land spreading. When digester effluent with 5.1×10^4 salmonella/ml was stored at ambient temperature for 24 days numbers were reduced to only 10 organisms/ml. Research by Taylor & Burrows (1971) indicated that such low numbers (less than about 100/ml) of salmonella would be unlikely to cause infection even in cattle allowed to graze on grass recently sprayed with digested slurry.

In the case of digested municipal sludges there has never been reported a case of disease transmission to human beings due to contact of employees in either waste-treatment plants or sludge-spreading operations.

Guidelines have been laid down to make spreading economical and safe. Spreading of digested sludge has been approved for the growth of all crops except those vegetables normally eaten uncooked and caution is advised against feeding digester sludge-grown forage for milk cows for approximately a two-month period after sludge spreading.

Digested sludge has been found (like the pig sludge) to be an effective source of N, P and micronutrients, but lacking in K. Higher applications tend to increase the nitrate nitrogen content of leachate waters and this appears to be the main factor which could limit the application rate. Experiments on the use of digested municipal sludge as fertiliser have been done and some are mentioned below. The feedstock sludges for municpal sewage digesters are generally of lower solids content (3–4 %) than those used (or proposed) in farm-waste digesters. Various degrees of dewatering and thickening of the digested sludge will occur in different sewage works, and, as said before, the analysis of the feedstock varies. So, as should be the case with all sludges, application rates should be based on the analysis of a particular sludge and the fertiliser status of the land. The following text should be read with this in mind, but the experiments give some general indications for the use of sludges.

While copper has been mentioned in the case of pig sludges, the possible presence of other heavy metals (e.g. chromium or zinc) and the dangers of these contaminating land and plants used as foodstuffs has been a topic of discussion for some time. Opinions about the dangers from these metals vary, as do their concentrations in different sludges. Present legislation requiring factories to purify waste-waters and specifying limits for pollutants, including metals, introduced into sewers should help to decrease the metal content of sewage sludges.

Lutrick & Bertrand (1976) conducted trials to determine the effect of digested sludge applied to soil for growth of grain sorghum, corn and soybeans. Also studied was the effect on the accumulation of nutrients, certain micro-organisms and heavy metals in the plants and in the soil.

Excessive rain during the trials may have caused aberrant results of plant growth due to possible leaching of nutrients. It was found that application rates of 7·5 cm, 15·0 cm and 22·5 cm all produced less corn than did similar applications of commercial fertiliser. The lower application of digested sludge produced less grain sorghum than commercial fertiliser whereas the two higher rates produced significantly more. For soybeans, all applications gave higher results than commercial fertiliser. It was also found that sludge produced one week earlier maturity of all crops.

Soil analyses showed that faecal coliform bacteria took up to four months to die off and that there was an increase in phosphorus, zinc and lead in the soil after application but a decrease in potassium. The increase of zinc in the soil and also in plant tissues suggested that this metal could reach toxic concentrations with repeated applications of sludge.

Work by Trout *et al.* (1976) to evaluate the environmental hazards of land application of digested sludge confirmed that faecal coliforms did not survive for long periods in the soil, in this case only 2·5 months, and, when present, did not extend more than 120–150 cm into the soil.

The main problem associated with sludge spreading was found to be the leaching of nitrate into the ground water. (Nitrate would be formed by oxidation by soil organisms of ammonia originally present in the sludge). This was substantial and increased with each sludge application. It would appear that this would be the first limiting factor for sludge application on a yearly basis.

An associated short-term problem was found to be that of salt building up in the soil, affecting plant growth. To remove this would require an application of water which, unless closely monitored, could cause further leaching of nitrate. Metal contamination of the ground water was not considered to be a problem and uptake by plants was not, in this study, enough to cause any toxicity.

The use of dried anaerobically digested sludge as fertiliser on agricultural soil and as a soil conditioner and fertiliser on acid strip-mine spoil was carried out on tall fescue and alfalfa grown in greenhouse plots (Stuckey & Newman, 1977).

An application of 314 tonnes of sludge/ha increased the yield of both tall fescue and alfalfa whereas an application rate of 627 tonnes/ha further increased the yield of tall fescue alone compared to untreated controls.

Applications to the acid strip-mine spoil greatly improved its quality, giving much greater plant yields over untreated controls. This was attributed to the increase in pH of the soil (3·5–5·5 for the 314 tonnes/ha application and 3·5–6·0 for the 627 tonnes/ha), the increased soil volume for root growth, and the availability of additional nutrients.

Increasing rates of digested sludge applications decreased the levels of manganese, zinc, nickel and cadmium accumulated in the crops in strip-mine spoils but did not affect copper accumulation. No plant toxicity symptoms were observed after two years.

A large proportion of the total nitrogen present in digested sludges is in the form of ammoniacal nitrogen. Much of this could be lost by volatilisation when a sufficiently large gradient in ammonia nitrogen occurs between the applied sludge and the atmosphere under drying conditions.

To evaluate this phenomenon Beauchamp *et al.* (1978) measured the volatilisation in May for five days and October for seven days after applications of digested municipal sludge of 116·48 tonnes/ha and 134·40 tonnes/ha respectively. Ammoniacal flux was found to follow a diurnal pattern with maxima occurring about mid-day. The flux decreased exponentially with time producing 'half-lives' of 3·6 and 5 days respectively. Air temperature appeared to have some effect on the flux rate, especially for the first two to three days after spreading.

It was estimated that during the five day experimental period in May, 60 % of the 150 kg ammonia nitrogen per hectare applied in the sludge was lost. Similarly, in October, 56 % of the 89 kg ammonia nitrogen per hectare applied was volatilised.

Digested piggery slurry can be converted to a slow-release fertiliser (Watson *et al.*, 1979) by combining the digested sludge with urea and formaldehyde in a cold process. The sludge, urea and formaldehyde mixture is acidified to pH 4, left 12 hours to semi-solidify to a granular consistency and finally neutralised with lime. The supernatant is run off and can be used for irrigation or, following a phosphorus removal, discharged into a water course. The remaining solids are then mixed with peat moss to form a stable, easily stored and handled, fertiliser. By variation of the reaction conditions the release properties of the product can be varied. With further work it was hoped to produce a fertiliser in which 50 % of the nitrogen would be released in six months. This type of fertiliser would be best fitted for horticultural application to gardens, golf courses, amenity areas, forestry and nursery stock.

USE IN ANIMAL FEEDSTUFFS

The feed value of digested sludges resides, as discussed previously, in the nitrogenous materials and to some extent in the residual fats. These are contained in the bacterial cells in the sludge. Sludge is thus a form of single-cell protein (SCP) diluted with fibrous material of little or no feed value.

Raw animal wastes have been used for some years for direct addition to carbohydrate-containing feeds such as grains or have been used in the preparation of a type of silage. The results have been variable with different animals, different levels of feeding and different types of waste. The use of digested farm-animal wastes as feed additives is thus a logical proposition. The digested wastes should obviate difficulties in handling caused by the smells of raw wastes, and the digested sludges should be more stable during processing and storage than the raw sludges. Vitamin B_{12} is known to be present in digested sludge and a commercial process for its production from municipal digesters was used, as mentioned earlier. Other B-vitamins of bacterial origin should also be present in greater amount in digested than in raw wastes. For full utilisation of bacterial protein nitrogen and ammonia nitrogen the sludges should be added to ruminant feeds. The amino acid composition of the digester bacteria will be similar to rumen bacteria (see, for instance, Hobson, 1969) which in turn are similar to other bacteria.

The question of possible transmission of pathogens in sludges used in feeds can cause some worries, but this problem may be less than that arising when raw wastes are used. Digested vegetation should, of course, be free from pathogens as the feedstock will have generally little, if any, contamination. There exists the possibility, though, that if use of digested sludges as feedstuffs became general they would have to conform to some bacteriological standard and so some form of pasteurisation or sterilisation by heat or chemicals would have to be introduced into the processing. The present position of raw animal wastes in feeds is rather unclear.

It is outside the scope of this book to discuss in detail the possible uses of digested sludges as feed additives but the following notes indicate the type of results that have been obtained in experiments on this topic. Apart from the use of the digested sludge directly, the use of the sludge or fractions of it as substrate for growth of algae or other micro-organisms and the use of these latter as animal feeds, is also possible. Digested municipal sludges have not much, if at all, been used as feeds, probably because of the views held by the public about the use of 'sewage' in production of anything for human consumption. But there is also, again, the question of heavy metal content of the sludges.

Some work has been done on the digestion of cattle-waste supernatant, with additional carbohydrate, as a source of animal feed. Reddy & Erdman (1977) supplemented the liquid portion of feedlot waste with waste carbohydrate in the form of molasses, whey, cornstarch or starch recovered from potato-processing wastes. Digestion was carried out in a 20 litre, stirred digester at 40 °C or 44 °C for 24 hours. The resulting digester effluent was then concentrated five-fold using an evaporator. The final product contained 20 % or more crude protein, although VFA content varied significantly, depending on the source of carbohydrate used as supplement.

The authors concluded that the process was simple and efficient, with no high equipment costs, no supplementation with growth factors and no inoculum needed. The medium need not be sterilised and no other aseptic conditions needed to be observed. The process was simple, took only 24 hours and was easily adaptable to existing industrial technology. The product was stable and not subject to microbial spoilage when stored at 25 °C for several months.

A method for the enrichment of cattle manure by anaerobic digestion for use as a feed supplement has been reported by Moore & Anthony (1970). By adjusting the pH of digestion once daily for three days, the apparent crude protein level increased from 17 % to 43 %. Palatability tests with lambs showed that rations containing digester effluent were equal to those with ammonium lactate.

Raw animal wastes have been used successfully for a number of years to produce wastelage; the waste being mixed with hay and fermented (Anthony, 1966). It was established that breeding ewes and beef cows could be successfully maintained on wastelage alone (Anthony, 1967).

Preliminary studies using digester effluents to produce wastelage have been carried out (Summers *et al.*, 1980). Wastelage was successfully prepared from a mixture of 13 % digested cattle slurry, 6 % molasses and 81 % chopped barley straw. After 11 weeks ensilage at ambient temperature, the mixture was sweet-smelling and compared favourably with grass silage. Wastelage prepared from a similar concentration of poultry slurry was unsuccessful. There was no rapid fall in pH and little lactic acid production.

Thomas & Evison (1976, personal communication) looked at ways of inhibiting the methanogenic phase of the digestion of pig waste to produce an effluent rich in VFA for the growth of yeast. Two systems were used, a semi-continuous digester and an anaerobic contact digester. Methanogenesis was inhibited in the semi-continuous system by overloading to the point at which enough acids were being produced to inhibit the methanogenic bacteria. Experiments were also carried out in which the pH of the system was lowered artificially. Loading the digester at 0.4–7 kg $COD/m^3/day$ at detention times between one and 20 days produced an effluent with up to 6000 mg VFA/litre but large quantities of sludge were also formed.

The contact digester was run at dilution rates and sludge recycle ratios aimed at the elimination of the methanogenic bacteria. A liquid retention time of from a few hours to several days gave VFA yields comparable to the semi-continuous system but with much less sludge waste and lower retention times.

Both effluents supported yeast growth. Yields of between 0.4 and 0.7 mg yeast/mg volatile acids utilised were obtained with *Candida utilis*.

Another system devised to produce yeast cells for animal feed, but also successful for the cultivation of fungi, has been proposed by Moo-Young and colleagues (Moo-Young *et al.*, 1978; Moo-Young & Daugulis, 1979). This system consists of an anaerobic digester, a chemical hydrolyser and an aerobic fermenter and utilises both farm manures and crop residues.

The digester was used to digest farm manures and crop residues (in the initial experiments cattle waste and wheat straw were used) to produce a digested sludge, rich in nitrogen and VFA, for use as a nitrogenous substrate for yeast cultivation. Digester gas was used to heat the system which was run at a 14-day detention at 39 °C. Unlike most digesters, it was operated to balance gas production against substrate production and not to maximise gas production.

Chemical hydrolysis was used to remove hemicellulose from crop wastes by treating with dilute acid and heating to 100 °C for one hour. The resulting liquid, rich in sugars, was then mixed with the digester effluent and used as feedstock for the aerobic fermenter. This was an air-lift or bubble-column type of reactor. Pre-sterilised air was used for the system but no other aseptic conditions were provided, since the low pH of the yeast fermentation (pH 4·5) should keep contamination at a low level. If contamination did occur, this could be washed out by decreasing the pH further to between 3·5 and 4·0. The yeast cells were harvested by centrifugation or flotation, some 10 % used as inoculum for the next batch and the rest used either directly as a yeast cream or further dried to obtain a product of 6–10 % moisture.

The whole system was designed to operate in a cyclic-batch-mode with a 24 hour cycle time. The digester was fed once daily and a corresponding amount of digester contents removed at the same time and used to prepare the yeast culture medium. The aerobic fermenter was also run on a daily basis, 90 % of the contents being removed after this time and the rest left as seed for the next batch. The sugar solution from the acid hydrolysis could be prepared daily, or once a week to conserve energy.

The system was shown to be technically feasible for the production of *Candida utilis* and also of *Chaetomium cellulolyticum*, a fungus which utilises cellulose and can be used as a protein source. The solids remaining from the acid-treated crop residues can be further upgraded to highly-digestible ruminant forage material by a mild caustic treatment.

A wide range of crop and animal residues was found to be suitable for treatment in this way. For example, barley straw and cornstover were used to produce the sugars needed for utilisation by the yeast. A mixed-manure feed was used for the digester, using appropriate dilutions of the raw wastes. This was found to have several advantages. The relatively rapid breakdown of starchy, non-ruminant manure was counteracted by the slower utilisation of the cellulose-containing cattle manure and a balance was achieved between the faster digestion of, say, poultry or pig manure and the occasionally sluggish performance of ruminant manure digesters. Another advantage was the widely mixed population of micro-organisms present in mixed manure systems.

From an economic point of view, the process was energy conserving but not energy self-sufficient (33 % of the energy requirements for the whole process were produced) and was able to generate protein at $0·16/lb to $0·12/lb respectively when no biogas credit and when a biogas credit was applied for a plant capacity of 1 tonne protein per day (1979 figures). Raw material and energy costs were the largest contributors to the overall production costs of SCP and improvements in the utilisation efficiency of these would improve the economic outlook of the process.

Various experimental farms have been set up in Taiwan (Chung Po *et al.*, 1975) to produce from digester effluent algae as a protein source for feeding to pigs and poultry. Although raw sludge could be used, the Taiwan authorities recommend that this first be digested as most intestinal parasites, their eggs and pathogenic bacteria are killed during the digestion process.

Effluent from digesters such as those previously described is run into shallow troughs (2 m × 4 m × about 25 cm deep) to allow light and air to penetrate as much as possible. A settled, digested-sludge liquid volume of about 1 % of the pond volume is suggested as the daily loading. Troughs may be arranged in a cascade formation with a motorised or wind- or water-wheel-driven pump to recirculate the dilute sludge. Alternatively, they may be all on the same level with water circulation by windmill-driven rotor.

Algae are harvested after four days whilst the cells are still young. Older cells tend to be indigestible. Drying the cells also tends to reduce digestibility so cells are fed as a wet sludge, mixed with other feeds. A strain of *Scenedesmus* has been found to give best results and this can be obtained from official sources. Once a pond has been started, enough algae are always left after harvesting to start a new growth.

Some work has also been carried out in California on combining algal ponds with an anaerobic digester treating poultry waste. It is doubtful, however, if algal production could ever be successful on a commercial scale due to the large areas of land required for ponds. Filtering or centrifuging systems for the harvest of the algae and a system for recirculation of water are also of high cost.

REFERENCES

ANTHONY, W. B. (1966). *Proc. Nat. Symp. Anim. Waste Manag.* ASAE Pub. No. AP-0366, p. 109.

ANTHONY, W. B. (1967). *J. Anim. Sci.* **26**, 217.

BEAUCHAMP, E. G., KIDD, G. E. & THURTELL, G. (1978). *J. Environ. Qual.* **7**, 141.

CHUNG PO, WANG, H. H., CHEN, S. K., HUNG, C. M. & CHANG, C. I. (1975). In: *Managing Livestock Wastes.* Proc. 3rd Int. Symp. Piggery Wastes. Am. Soc. Agric. Eng. p. 238.

GRACEY, H. I., STEWART, T. A., WOODSIDE, J. D. & THOMPSON, R. H. (1976). *J. Agric. Sci. Camb.* **87**, 617.

HOBSON, P. N. (1969). *Proc. Biochem.* **4**, 53.

HOBSON, P. N., BOUSFIELD, S. & SUMMERS, R. (1974). *Crit. Rev. Environ. Control* **4**, 131.

HOBSON, P. N., BOUSFIELD, S., SUMMERS, R. & MILLS, P. J. (1980). In: *Anaerobic Digestion* (eds. D. A. Stafford, B. I. Wheatley and D. E. Hughes) Proc. 1st Int. Symp. An. Dig., Cardiff. Applied Science Publishers, London.

HOBSON, P. N. & SHAW, B. G. (1973). *Water Res.* **7**, 437.

LUTRICK, M. C. & BERTRAND, J. E. (1976). *Proc. Inst. Environ. Sci.*, 22nd Ann. Tech. Meet., p. 528.

MOORE, J. D. & ANTHONY, W. B. (1970). *J. Anim. Sci.* **30**, 324.

MOO-YOUNG, M. & DAUGULIS, A. J. (1979). In: *Ind. Waste Treatment and Utilisation*, Pergamon Press (In press).

MOO-YOUNG, M., MOREIRA, A. R., DAUGULIS, A. J. & ROBINSON, C. W. (1978). *Biotech. Bioeng. Symp.* No. 8, p. 205.

PFEFFER, J. T. (1973). Final Rept. No. EPA-R-80076 U.S. Environmental Protection Agency, Nat. Environ. Centre, Cincinnati.

REDDY, A. C. & ERDMAN, M. D. (1977). *Biotech. Bioeng. Symp.* **7**, 11.

SHAW, B. G. (1971). Ph.D. Thesis, University of Aberdeen.

STUCKEY, D. J. & NEWMAN, T. S. (1977). *J. Environ. Qual.* **6**, 271.

SUMMERS, R. & BOUSFIELD, S. (1976). *Proc. Biochem.* **11**, 3.

SUMMERS, R. & BOUSFIELD, S. (1978). In: *Proc. Seminar Anaerobic Digestion of Farm Wastes*, ADAS, Reading.

SUMMERS, R. & BOUSFIELD, S. (1980). *Ag. Wastes* **2**, 61.

SUMMERS, R., BOUSFIELD, S. & HOBSON, P. N. (1980). In: *Anaerobic Digestion* (eds. D. A. Stafford, B. I. Wheatley and D. E. Hughes) Proc. 1st Int. Symp. An. Dig., Cardiff. Applied Science Publishers, London.

TAYLOR, R. J. & BURROWS, M. R. (1971). *Br. Vet. J.* **127**, 536.

TROUT, T. J., SMITH, J. L. & MCWHORTER, D. B. (1976). *Trans. ASAE*, p. 266.

WARD, R. L. & ASHLEY, C. S. (1977). *Appl. Environ. Microbiol.* **34**, 681.

WARD, R. L., ASHLEY, C. S. & MOSELEY, R. H. (1976). *Appl. Environ. Microbiol.* **32**, 339.

WATSON, J., O'SHEA, J., SPILLANE, T. A. & CONNOLLY, J. F. (1979). *Conserv. and Recycling* **2**, 269.

CHAPTER 7

Biogas Production—Laboratory and Pilot-plant Experiments

An essential preliminary to building of large-scale digester plant is small-scale tests to determine whether the proposed feedstock digests and what are the optimum conditions for digestion.

Small-scale tests are also an essential preliminary in assessing the feasibility of a particular digester design (e.g. single-stage, two-stage, two-phase and tower) for digestion of a particular waste, or a particular type of waste, before large-scale construction is embarked upon.

Small-scale tests can be used for determination of the effects of various additives on digestion and the effects of changing the concentration of some intermediate or final product of the reactions. Laboratory-scale apparatus can also be used for studying the effects of deviations in running parameters on steady-state conditions in a digestion. Such apparatus can be used for some kinds of 'trouble-shooting' in the case of failure of a full-size digester, and to find methods of restoring the digestion.

In all these cases what is being used is, in essence, a small-scale digester, with a full digester flora. But the laboratory experiments may be carried a step further in that individual types of bacteria or groups of bacteria may be cultured and the effects on these of changing conditions, inhibitors or activators, etc., may be observed, so ascribing the effects to one particular part of the complex pathways of digestion.

Some of the uses of bacterial cultures and of whole digester contents in small-scale experiments have been mentioned in previous sections. What is described here is mainly the use of small-scale apparatus in determining the biodegradability of various possible digester feedstocks.

GENERAL—THE APPARATUS

How small is 'small-scale'? This depends on what is being investigated, but one of the biggest factors in deciding the scale of apparatus in investigating anaerobic digestion is the feedstocks used. These are almost always suspensions of particulate matter, and very often thick slurries containing large particles. Engineering problems cannot, in general, be investigated on a small scale. While test heat-exchangers or stirrer systems, say, can be made on a small scale and of the materials to be finally used, or some other material if, for instance, flow inside a vessel or pipe is to be seen, such apparatus can only be used with water or very thin, fine-particle, aqueous suspensions. When large-particle slurries are to be tested there is obviously a minimum size of pipe or chamber through which these can flow, so one inevitably begins to think of the 10-, 20- or 30-m^3 size of digester for engineering tests. But this is still of small scale compared with the full-size digesters of hundreds of cubic metres capacity, and some problems only become apparent and can be investigated by building on this full-size scale.

The small unit used for biological experiments cannot, then, reproduce the mechanics and materials of the large unit. But it can reproduce the mode of operation of the full-size plant. However, even here the feedstock imposes limitations. With a substrate in true solution the physiology and kinetics of anaerobic bacteria growing at doubling times measured in hours can be investigated in 300 ml continuous cultures (e.g. Hobson & Smith, 1963; Hobson, 1965a, b; Hobson & Summers, 1967). If doubling times of many days are to be investigated then the very small hourly volumes of medium required for a 300 ml culture cannot be added in a continuous stream. The culture vessel must be increased in size so that the small pumps can add a reasonable hourly volume of material and work continuously. But even if the vessel is 5-litre or so, a 20 day detention time means that only some 250 ml of feedstock has to be added each day.

Small flow-rate, laboratory-scale pumps are of piston, diaphragm or most often nowadays peristaltic, type. To pump small volumes they need small-bore tubing and this cannot take large particles. The authors also found that even with larger, roller-type peristaltic pumps, gelatinous and fibrous material such as animal-waste slurries tended to get squashed into lumps by the rollers and blocked the tubes. A finger-type peristaltic pump was used successfully in continuous operation over many months feeding a digested piggery slurry of 2–3 % TS to a small trickling filter, but the digested slurry has properties different from those of the raw slurry.

The bench-scale continuous digester, then, is either hand-loaded by pouring in the feedstock in some way or is loaded in a modification of this method by some timed valve operation which allows a certain amount of feedstock to fall through a wide-bore tube from a reservoir or something of this kind. But this kind of apparatus cannot be run at anywhere near a continuous flow.

A further problem that can occur with any kind of mechanised, small-scale feedstock addition is that of keeping the feedstock properly mixed and of adding a properly representative suspension in each dose. If small amounts are to be added manually, at intervals, there can, again, be difficulties in taking each time the same representative sample from the larger feedstock reservoir.

Such considerations make about 5 litres a minimum volume for any kind of 'continuous-flow' digester system. The authors and colleagues, after some tests, decided on 15 litres as a suitable scale for laboratory digesters. This allowed mechanical stirring, but not continuous mechanical feeding. The piggery-waste feedstock was added once a day by a hand-operated, suction system. This, it was realised, was not continuous operation, but the volumes added (0·5–1 litre or so) were sufficiently large to get a representative sample of the feedstock slurry (Shaw, 1971; Hobson & Shaw, 1971).

To get a digester system that could be mechanised and with an input which could be added at frequent intervals the authors decided on a 100-litre digester and later added 150-litre digesters (Bousfield *et al.*, 1972, 1979). Even with the first, 100-litre digester the scroll-in-stator pump would not pass pig slurry of more than about 4–4·5 % TS without danger of blockage. For higher TS experiments very thick slurry had to be added by hand at intervals while the main volume of feedstock was added by the pump. The pumps fitted to the 150-litre digesters are linear-flow peristaltic pumps and can move slurries of 12 % or more TS.

It can be suggested that some of the problems of small-scale work could be overcome by grinding the feedstock slurry to small particle size. However, the authors have always looked upon their experiments as providing data for full-scale systems and reducing the particle size of feedstock could alter the kinetics, and the extent, of breakdown of the material. Similarly, removing any large amount of big particles can alter the results.

With a batch digester a smaller experimental system may be suitable as the digester has only to be loaded once, and may not even need to be stirred. One or two litres could be big enough.

A further important point about laboratory digesters is that the whole system must be made in such a way that it can, if it is a continuous-flow system, run for months without being stopped for more than the odd hour or two for repairs. And it should not need to be emptied and cleared out, and so on, during a run. With a continuous-culture system a period of two or possibly more detention times is needed after a change in running parameters (temperature, loading rate, etc.) for the bacteria to again settle down into a steady state. With 20- or 30-day detention times this is, obviously, $1\frac{1}{2}$–2 months. The initial stabilisation of the digester flora may take some weeks after inoculation. A microbial system such as a digester, with a very complex bacterial population and a complex feedstock that sometimes cannot be completely controlled in composition, will not attain the almost unvarying steady-state of a pure bacterial culture on a simple, dissolved substrate. Some oscillations are found to occur. So a long running time and a large number of determinations are needed to get good average results. In addition, with a system of slow turnover, inhibitions or adaptations in the bacteria may take a long time to develop. The authors have noted declining production in some digestions with high loadings only after some months' running. Experiments with laboratory-scale apparatus that run for only days or a few weeks should be viewed with caution, except, of course, in the case of batch culture.

A full-scale digester is, hopefully, going to run continuously for many years. The laboratory-scale tests should, then, be long enough to at least show that it will not fail after a few months. Another reason for long-term tests could be the following. Only long-term tests can show whether the bacterial population will be stable for long periods, or whether it might need reinoculation. This latter is *a priori* unlikely in the case of excreta feedstock as this itself will reinoculate the digestion, and experiment shows this to be the case. The authors' 100-litre piggery-waste digester has been running for about 10 years without reinoculation and cattle- and poultry-waste digestions have been running for three years. However, it is possible that mutations and death of bacteria could result in digestions of factory wastes or 'energy-crops' requiring periodic reinoculation as these feedstocks contain few, if any, of the digester bacteria.

Laboratory-scale digesters have been made of various materials. Some have been conventional laboratory bacterial culture vessels, but in general the sophistication of such apparatus, which is designed for sterile operation, is not required. Others have been laboratory constructions in glass or plastic or metal, but, obviously, the metal should not be copper or brass or zinc-plate or other potentially 'attackable' and toxic material. Brass in gas lines, gas taps or meters corrodes, with sulphide formation. The first of the 15-litre digesters mentioned above was actually adapted from a stainless-steel milking-machine vessel. The second two were made from $\frac{1}{4}$ in opaque plastic sheet and plastic drain-pipe sections, welded together where required. Stirring was by paddles on a shaft driven by a geared motor bolted to the top of the vessel, the shaft passing through a mechanically-sealed, gas-tight top bearing and running in a bottom, steady, bearing. The second two ran more or less continuously for three years until 1971.

If glass or clear plastic is used a cautionary note is sounded by the work of Torien (1967) who found that photosynthetic bacteria could grow in transparent vessels.

Small vessels are generally heated by immersion in a water bath, although some laboratories with the facilities keep small-scale digesters in a hot room. One must beware of growth of micro-organisms in feedstock tanks in such cases.

Stirring is generally by low-speed paddles as mentioned above, although some digester vessels have been placed on mechanical shakers, and some are just shaken by hand at intervals. Gas-stirring could in some cases, be used, the gas being supplied as carbon dioxide or nitrogen from a high-pressure cylinder with suitable reducing valves. However, such stirring would make registration of evolved gas very difficult.

Apart from entry and exit ports for feedstock, provision must be made for sampling the digester contents. Determinations of solids, COD and BOD are best done on large samples collected from the overflow to avoid difficulties in withdrawing representative samples by pipette or other means. Determinations of substances in solution or in the small-particulate bacterial fractions such as volatile acids or ammonia can be done reasonably well on small samples withdrawn from the digester. The authors use a system similar to that used in bacterial culture apparatus. A tube, about $\frac{1}{4}$ in i.d. has one end half-way down the digester liquid, and the other end, outside, is connected through the lid of a screw-cap bottle ('Universal' type, 25 ml) which has a rubber liner. The liner is exposed by another hole cut in the metal top, and this has a wide-bore hypodermic needle fixed through it. A clean bottle is attached to the top and a 50 ml hypodermic syringe to the needle. A vacuum created in the bottle by the syringe sucks up liquid from the digester. A few pushes and pulls on the syringe usually suffices to clear any blockage in the tube.

Measurements of pH can be done on samples taken like this, although for exact results the sample should be allowed to re-equilibrate with carbon dioxide at digester temperature before measurements are made. The authors have had no success with pH electrodes in the digester. Glass electrodes and porous plugs of reference electrodes quickly become poisoned and choked with sulphides and layers of proteins, fats, etc.

Evolved gas has been measured by water or salt-solution displacement in graduated cylinders and gas burettes. With the 15-litre digesters mentioned above a 'diving-bell' collector working a mechanical counter and similar to

Fig. 7.1.   Pilot-plant 150-litre digesters in Aberdeen, used to determine optimum conditions for digestion of farm and other wastes. The digesters can be run as single-stage, two-stage or feedback, stirred-tank systems.

that first described by Merkens (1962) was used. Small wet, or dry, gas meters capable of registering a few litres a day are also commercially available, although some of these are now very expensive, in Britain at least. Both types have been used on the authors' 100- and 150-litre digesters. Condensate pots may need to be placed in gas lines before meters.

Methods of analysis will not be described here as almost everyone uses the standard sewage-works methods for solids—BOD, COD, ammonia, etc. (e.g. *Standard methods*, 1965). The volatile fatty acids in the digester or effluent are by convention determined as acid equivalents and reported as mg acetic acid/litre. The actual VFA present can be determined by gas-chromatography by various published methods. Gas analysis is always difficult. The authors have always used a small mass-spectrometer, partly because one was available in the laboratory, but they have more recently also used gas-chromatography; again methods are published. Some other types of commercial gas analysis apparatus may also be suitable, and less costly, but one should first make sure that they can analyse in the 50–100 % methane range and are not just suitable for detecting low concentrations.

The 100-litre and above experimental digesters are really small pilot-plant and as such are made in metal. The authors' plant (Fig. 7.1) is constructed of stainless-steel to industrial fermenter standards. It was made thus so that it can be used for sterile fermentations other than anaerobic digestion. If only digester work is to be done then simpler plant can be built, but attention must be paid to provision of gas-tight bearings for stirrers and so on. The plant in Fig. 7.1 has gas-tight overflow tanks and so no weir system is needed to maintain the digesters air-tight and at 6 in WG pressure set by the water-sealed gas holder to which they are connected. The overflow tanks are fitted with stirrers to ensure that no solids are left behind when they are emptied, through wide-bore ball valves, for analysis of the contents. The stirrers are operated for a few minutes before the tanks are emptied, and while the valve is open a stream of carbon dioxide is passed into the tank to replace the issuing liquid and ensure that air is not sucked in through the valve. The digester tanks are stirred by variable-speed paddle stirrers, and are heated or cooled by thermostatically-controlled water jackets. The feedstock tanks are continuously stirred to ensure homo-geneous feed, as the pumps work on a short time interval (5 to about 15 or sometimes 30 min) between 'on' periods. However, as these small pumps cannot be guaranteed not to make some selection in particle size over a long period, the pipes from the pumps to the digesters are arranged so that they can be disconnected and the input directed into sampling flasks (the pumps are run continuously for a few minutes to provide a litre or more of sample for analysis).

In the much larger, 13-m^3 small-farm-scale digesters samples of feedstock are taken from the stirred, input tanks as the larger pumps make no selection. Output samples are taken from the overflow at a time when overflow occurs naturally and samples from the digester can also be taken from valves on the heater sludge-circulation pipes.

Detailed descriptions of small-scale digesters have not been given as there have been so many kinds, but most of the papers quoted in the following sections give some description of the apparatus used.

DOMESTIC SEWAGE

Since full-scale sewage digesters have been running for many years, there has been little need to test the feasibility of mesophilic sewage digestion on a small scale. The work that has been done has been directed towards improving methane production, the possible toxic effects of substances likely to be present in the sewage, the feasibility of thermophilic digestion, and 'trouble-shooting'.

Looking for improvements in methane production Haug *et al.* (1978) studied the effect of thermal pretreatment on the digestibility and final dewatering of sludges. Treatment of activated sludge at 175 °C for 30 min in a pressure vessel before digestion, produced a 60 % increase in methane production and a 36 % decrease in Volatile Suspended Solids (VSS) in the sludge. However, problems of toxicity could occur unless the digester bacteria were given time to acclimatise to the heated sludge. Dewaterability of the digested sludge was increased by heat treatment. Similar pretreatment of primary sludge did not increase digestibility but did appear to increase dewaterability of the digested sludge. The authors pointed out that one of the problems in thermal treatment of sludge was the odiferous compounds produced, but that these were significantly reduced during digestion. The authors calculated that thermal pretreatment prior to digestion may result in an increase in net energy production. One of the difficulties in large-scale treatment of sludge is fouling of the heat-exchangers, but the authors suggested that this was a matter of design and construction. For their small-scale work they used either a glass-lined or a Monel-metal pressure vessel.

The effects on digestion of chemical treatment of sludge were investigated by Gossett *et al.* (1978). They studied pilot-plant digestion of domestic sewage dosed with alum, ferric chloride or organic coagulant and compared results with the digestion of the untreated sludge. In all treatments, gas production, methane content of the gas, and COD and VSS reductions were all lower than the control. Dosing with 200 ppm of alum reduced digestibility to 92 % of the control and 400 ppm to 30 %. Ferric chloride at 100 ppm reduced digestibility to 90 % and 200 ppm to 78 %. Dosing with up to 300 ppm of organic coagulant had no effect, but 475 ppm reduced digestibility to 81 %.

The problem of heavy metals in digestions have been previously mentioned, and these have been investigated in the laboratory for many years. Early work on the effects of metals was carried out by Rudolphs & Zeller (1932) who found that heavy metal ions were more toxic than light metal ions in digestion of domestic sludges.

Barth (1967) summarised the results of a 10-year study of continuous- and pulse-feeding of heavy metals. They showed that a significant amount of heavy metals were removed from solution during digestion. The metals tend to accumulate in the sludges produced in sewage treatment. In this work removals of up to 60 % of the metals in solution in the sewage by concentration in primary and secondary sludges resulted in increase in metal concentrations in the digester feed sludge of six to thirty-seven times that in the sewage-plant influent. Zinc, chromium, copper and nickel salts were used.

Similar effects have been observed with mercury, cadmium, lead, manganese and iron (Chen, 1974; Cheng, 1975; Davis & Jacknow, 1975; Ghosh & Zugger, 1973; Neufeld & Hermann, 1975).

As previously mentioned, heavy metals can be tolerated in digestion processes if they are precipitated as sulphides. Lawrence & McCarty (1965) showed that little effect on digestion was produced by feeding sulphides of copper, zinc, nickel and iron to laboratory digesters, while metal chlorides produced a marked reduction in performance. Ghosh (1972) later confirmed the effects of sulphide.

Hayes & Theis (1978) showed that heavy-metal inhibition reduced both gas production and the proportion of methane in the gas, and so volatile acids accumulated. Shock loadings of heavy metals inhibited all bacteria. Toxicity of the metals studied decreased in the order: nickel, copper, lead, chromium, zinc, with cadmium having no toxic effect. Toxicity coincided with near maximum uptake of the metals into the biomass, and could be prevented by the addition of precipitating ligands such as sulphide, while operating the digester at near maximum pH.

A novel way of reducing the effects of toxic materials in domestic digesters was described by Koch *et al.* (1978). They found that the addition of 2–5 % powdered, activated charcoal to batch-loaded, pilot-plant digesters improved the performance and stability when the system was overloaded.

Bonomo (1975, *a, b.*, reviewed by Ghosh, 1972) conducted pilot-plant, sewage-sludge digestion studies which showed that inhibition by iron began at between 300 and 400 mg/litre, was significant at 500 to 1000 mg/litre and total between 1500 and 2000 mg/litre.

Work has been carried out on the effects on digestion of primary sludge of vegetable oil and alkyl benzene sulphonate (ABS) which was used in domestic sewage works as an emulsifier. This work was promoted by the fact that many sewage works operators have, or had, reservations about adding grease to their digestion tanks. It was found that cottonseed oil in its natural state was easily digested with primary sludge, producing approximately 1290 ml of gas per gram of oil added, whereas the addition to the oil of 95 ppm ABS resulted in a serious decrease in gas production, solids destruction and pH. (Johnson & Bloodgood, 1958). The general effects of detergents on digestion have previously been discussed.

Considerable work has been carried out on the possibilities of thermophilic digestion of domestic sewage. Amongst the earliest Rudolphs & Heukelekian (1930) and Heukelekian (1930) found that the yield of gas per gram of volatile matter was high and that a greater proportion of the volatile matter was destroyed than in a similar mesophilic digestion and that detention times could be shorter. There was a slight reduction in fat degradation when compared to mesophilic digestion. Gasification in batch digesters was complete after 14 days at 50 °C, 11–12 days at 55 °C and 60 °C and retarded at temperatures above 60 °C.

An optimum temperature of 50 °C for batch digestion of primary and activated sludge was reported by Fair & Moore (1932; 1937). No lasting effect occurred when the temperature was dropped to 40 °C and 20 °C and then returned to 50 °C (Heukelekian & Kaplowsky, 1948). However, Golueke (1958) observed that when a digester bacterial population was not well established it became very sensitive to any abrupt drop in temperature with the result that there followed a decline in destruction of volatile matter. He also reported that sludge produced by digestion at 50 °C and 60 °C had a much improved dewaterability.

In a review paper on the thermophilic digestion of sewage sludge Buhr & Andrews (1977) concluded that full-scale work in Los Angeles (USA) and Moscow (USSR) (see later section) had shown that thermophilic digestion was possible but it had some negative as well as positive results. The points for and against thermophilic digestion are summarised later when large-scale work is described. However, Buhr & Andrews proposed a dynamic model to give a qualitative description of behaviour of the process under varying operating conditions and found that the model predicted that process failure could be caused by sudden temperature changes and also indicated that there is an optimum operating temperature for maximum volatile acids removal and maximum stability at a given detention time. The optimum temperature increased as the detention time was decreased.

Before the previously mentioned pilot-plant for domestic sewage digestion by a type of sludge retention digester and anaerobic filter was investigated by Pretorius (1971) tests were made with an 8-litre digester followed by a filter of similar volume filled with stones of 2 cm diameter at the bottom graded to filter sand at the top. Thirty five to forty percent of Suspended Solids were hydrolysed in the digester while most of the gas production occurred in the filter.

Lettinga *et al.* (1980) are experimenting with the sludge-blanket upflow digester for treatment of municipal sewage waters, using laboratory- and pilot-plant.

An example of 'trouble-shooting' using laboratory apparatus is given by the tracing of the previously mentioned inhibition of full-scale domestic sewage digesters by pentachlorophenol (Drew & Swanwick, 1962). After the drop in gas production in the digesters was noted, 3-litre laboratory digesters (Swanwick *et al.*, 1961) were used to determine whether the digestion of primary sludge was being influenced by anything accumulating in the surplus activated sludge. The results were inconclusive. However, further tests were then carried out on digestion of sludges collected from each of the three main sewers serving the sewage works. These tests showed that sewage from one catchment area was causing the inhibition. Further small-scale tests on sewage collected at various points showed that one branch of this sewer was that containing the inhibitory sewage. The source was then further pin-pointed by tests on discharges from a number of factories in the area of the branch sewer. The factory wastes were digested with a sludge known to come only from domestic sources, and in the proportion in which the factory waste was added to the main sewage flow. These tests showed that one factory was responsible for the inhibitory discharge and after further tests to determine the source of the pentachlorophenol in the factory (only a part of the waste discharge contained the chemical) and to determine inhibitory concentrations, steps to eliminate the inhibitor were taken and the large-scale digesters were recommissioned.

Sanders *et al.* (1979) used 4-litre stirred flasks with a working liquid volume of 2 litres to investigate the fate of poliovirus in domestic sewage digestion. Since the free virus in sewage waters becomes incorporated in the activated sludge of the aerobic treatment plant and so transferred to the digesters, in the experiments, virus was incorporated into activated sludge and this was used as the feed for the digester flasks. The feed was introduced to the digesters each day and an equal quantity of digester contents withdrawn. The digesters were run at 34°, 37° and 50°C and at five- and ten-day detention times. The experiments were detailed, but, briefly summarised, showed that the virus was deactivated and that deactivation was related to temperature. At 50°C temperature was the only factor involved in deactivation as the latter occurred so rapidly; at mesophilic temperature factors other than heat were involved. In the mesophilic range recovery of active virus was inversely proportional to the Volatile Solids (VS) concentration in the digester and detention time had a small, but direct, effect at mesophilic temperatures, the greatest effect being at 34°C.

DOMESTIC GARBAGE

Domestic garbage (refuse) represents a very large source of substrate for energy production. Burning and pyrolysis are not always feasible as the dry waste of the household (papers, etc.) is mixed with wet vegetable matter and other kitchen scraps. Energy production from garbage is at present being investigated as, apart from considerations of conserving energy, disposal of the refuse by the conventional methods of landfilling is becoming increasingly expensive and difficult. Some cities in Britain now have to send refuse to areas far outside the city and vicinity to find suitable sites for dumps. Any process which can produce usable energy and also reduce the amount of material to be dumped while making it less polluting in the dump is worth research and development.

Domestic refuse, once the metal, glass and grit have been removed, is largely cellulosic fibrous material and plastics. The plastics can be burnt or pyrolysed, but many are virtually non-biodegradable. When the plastics are removed a largely cellulosic waste is left which tends to be too low in nitrogen to sustain good microbial growth. Thus experiments on digestion have usually been done after mixing the refuse with some other waste or cheap chemical high in nitrogen.

As described later, one, at least, successful large-scale digestion of sewage sludge and garbage was carried out some years ago, but laboratory-scale experiments have more recently been intensified to find the optimum conditions for garbage digestion.

To determine the digestibility of mixtures of garbage and sewage Klein (1972) tested shredded newspaper, paper pulp (Kraft pulp) and garden debris, in laboratory digesters, in various combinations with active domestic-digester sludge and then operated a pilot-plant, stirred digester of 317 gallons (1439 litres) capacity for six months. After seeding with digester sludge and bringing it up to full loading rate, the digester was operated on domestic sewage alone, at a 30-day detention time and 37 °C, for one month, loading three times per week. The sewage sludge was then gradually replaced by domestic refuse, 5 % at each loading until 50 % refuse and 50 % sludge were being fed. The metals and about 50 % of the glass had been removed from the refuse and the remainder had been ground to a maximum of $\frac{3}{4}$ in (1·9 cm) particle size. This 50:50 loading was continued for 10 weeks and then for the final four weeks 57·7 % refuse was fed. Loading was three times per week with 7·74 lb (3·52 kg) of air-dried garbage combined with 10 gallons (45·4 litres) of raw sewage sludge and 14·7 gallons (66·7 litres) of water. The Total Solids concentration would be probably 5–7 %. Sodium bicarbonate had to be added to keep the correct alkalinity, but this may not have been due to the garbage. Klein concluded that a 1:1 mixture of garbage and sewage sludge would digest. The Total Solids destruction was only 39·8 % because of the large inorganic content of the garbage, but the Volatile Solids destruction was calculated as 66·8 %. However, a problem which could indicate difficulties in a large-scale plant was that garbage contains a large proportion of floatable material and this material formed a thick scum layer in the pilot digester and remained largely undigested. This accumulation put the apparent solids destruction at 82·8 %. The scum layer had built up to a 1 ft thickness in a liquid depth of only 6 ft when experiments were terminated, and it is obvious that such an accumulation could not have continued long without digester failure.

The gas production was rather lower than that from normal sludge digestion and in the first three months of digester operation the methane content was only around 50 %, although in the last month it increased to 60 %. Klein suggested that the digestion of domestic garbage would be economical provided the cost of dewatering the sludge was reasonable, but obviously, such a small-scale study leaves many problems unanswered or unposed.

In 1975 Cooney & Wise published laboratory studies on garbage digestion using 50-litre stirred digesters, operated at a 30-day detention time with once-daily feeding. They used a mixture of 90 % shredded solid refuse and 10 % domestic sewage sludge (by weight). This was suspended in water to give a slurry of 2·5–5 % TS. The shredded refuse contained only 35 % VS and had a moisture content of 7 %. The sewage sludge contained 4·8 % TS of which 80 % was VS.

A thermophilic bacterial population for the digesters was developed by incubating various potential sources of bacteria (sewage, rumen contents, mud, etc.) at different temperatures up to 65 °C and adding these to digesters which were gradually increased in temperature up to 65 °C. The digestion gave better gas production than the mesophilic digestion, as at 65 °C, 11 ft^3 of gas was produced per lb VS added compared to 7·5 ft^3 of gas produced at 35 °C (687 and 468 litres/kg respectively), although the gases were not analysed. The digestion at 65 °C also operated satisfactorily with a volatile acid concentration in the digester of 1300 mg/litre. This was four times the normal concentration in the mesophilic digester.

Advantages of thermophilic garbage digestion were suggested as greater conversion of solids to gas, a better separation of sludge solids after digestion (although no experimental data are given) and better destruction of pathogenic bacteria and viruses (on the basis of results in the literature).

A disadvantage would be the high VFA content of the digested liquid which would increase the pollutional load to be disposed of by aerobic methods.

Pfeffer and his coworkers have probably done the most detailed small-scale experimentation and calculations on digestion of garbage plus sewage sludge. After initial laboratory work, pilot-plant experiments with a 100 US gallon (378 litre) digester, heated by a water jacket and stirred by a paddle operated semi-continuously at 36 rpm, were carried out. The digester was fed one volume daily through an opening hatch in the digester top after the withdrawal of an equal volume of digester contents.

Detailed results of these experiments are given in reports to the NSF (Pfeffer & Liebman, 1974) and in papers by Pfeffer (1973, 1980). Only a brief summary is given here. The small-scale results were used in the design of a full-scale plant referred to in a later section.

Ammonium chloride was added to increase the nitrogen content of the garbage. The feedstock used was 5–6 lb (2·26–2·72 kg) per day of shredded refuse (depending on its VS content) with 107·5 g ammonium chloride, 20 g potassium dihydrogen phosphate, up to 10 g $Cu(OH)_2$, 75–150 g Na_2CO_3, 2·5 litres fresh sewage sludge (about 3 % TS), with raw sewage (10 US gallons, 37·8 litres) as diluent. A thermophilic bacterial culture was developed and the digester run at 60 °C with a 10 day detention time.

Gas production varied with the VS content of the garbage from 6·28 ft³/lb VS added (390 litres/kg) at 90 % VS to 4·8 ft³/lb (300 litres/kg) at 60–70 % VS. The effects of digester temperature and other operating parameters were also studied.

Detailed studies on dewatering of the digested sludge, incineration and general methods of handling were carried out and power requirements calculated. The results were put into a model system which showed that the cost of gas production, if optimum laboratory conditions could be attained in practice, would be 8–10 cents (Canadian) per 1000 ft³ (28·3 m³). If the gas could be sold at 75 cents/million Btu, then, counting separating and shredding costs for the garbage, Pfeffer (1973) calculated that a net profit of 36–72 cents per ton of garbage could be made. However, whether such financial results could be achieved in practice is difficult to say and for the moment it would seem better to consider the process in the same light as is done for sewage digestion, namely as part of the garbage treatment system which produces a product better than the original garbage for landfill disposal and at the same time energy of use in the garbage plant or surroundings.

Work on thermochemical treatment of previously digested garbage was reported by Gossett & McCarty in 1975. They found that treating the garbage with acid or alkaline solutions at 133 °C for 180 minutes increased biodegradability and hence gas production significantly at both very low and very high pH but had little effect at more neutral pH. Under high pH conditions 40 % of the digested refuse was solubilised and about 13 % of refractory organics made biodegradable. At very low pH similar results were obtained.

In further work carried out in the same laboratory McCarty *et al.* (1976) found that solubilisation of organics could be increased by up to 90 % when the treatment temperature was raised to 250 °C at pH 13. It was found that a two-stage digestion system with heat treatment of sludge from the first stage at pH 13 and 200 °C resulted in an additional 73 % VS reduction and 73 % methane production. Inhibition of digestion did, however, occur when 1–3 g heat-treated lignin/litre was present and semi-continuous digestion required concentrations within or below these levels. Studies on the anaerobic breakdown of the single-ring compounds likely to result from the heat treatment of domestic refuse showed that, with acclimatisation of the bacteria, all were biodegradable and converted to methane by the mixed digester culture.

More recently Pfeffer & Khan (1976) have been studying the effect of a heat and caustic pretreatment on gas production in the anaerobic digestion of municipal refuse.

Using completely-mixed anaerobic reactors with a working volume of 15 litres, gas production was measured from the digestion of a daily feedstock addition of 100 g shredded refuse, 3·95 g ammonium chloride and 0·75 g potassium dihydrogen phosphate diluted to 2 litres with raw sewage. Two litres of digester contents were withdrawn each day and replaced by the feedstock. This gave a 7·5-day detention time. The digestion temperature was 60 °C. The rate of gas production was then compared with that of similar digesters fed on the same feedstock but in which the refuse had been subjected to a hot alkaline treatment. One hundred grams of refuse were mixed with 300 ml of an alkali solution and then heated in an autoclave for one hour. After cooling to room temperature, this material was mixed with the same quantities of chloride, phosphate and sewage as the control. Different concentrations of alkali solution and temperatures were investigated. Results showed that a pretreatment temperature of 130 °C and 3 g sodium hydroxide per 100 g dry solids gave highest gas yield. The increase in gas yield was about 20 % over the control. The treatment also increased the rate of gas production and the authors concluded that a high conversion efficiency was possible at much shorter detention times with pretreatment than without.

INDUSTRIAL WASTES

Anaerobic digestion has been used successfully for large-scale treatment of waste-waters from meat-processing plants, as is described later. Because of the volume of the wastes and the low solids content a feedback (contact) digester system was used. Pilot-plant experiments preceeded the large-scale plant.

Steffen (1958) found that gas from digestion of meat-processing waste contained 85 % methane and 11 % carbon dioxide. This is probably due to the high fat content of the waste. A vacuum degasifier was used to help to settle the bacterial sludge for return to the digester (see earlier section) and because of release of carbon dioxide from solution the gas from the degasifier contained 63 % carbon dioxide and 36 % methane. However, collection of this gas with the main stream from the digester tank would be worthwhile as Steffen found that up to 20 % of the total gas could come from the degasifier. Hemens & Shurben (1959) also operated a pilot contact-digester plant for meat-processing waste and found 80 % methane in the digester gas. Steffen found that $20 \, \text{ft}^3$ gas/lb VS destroyed ($1 \cdot 25 \, \text{m}^3/\text{kg VS}$) was produced whereas from domestic sewage $12–14 \, \text{ft}^3/\text{lb}$ ($0 \cdot 75–0 \cdot 87 \, \text{m}^3/\text{kg}$) is generally produced.

Wastes from fruit and vegetable processing are high in organic matter and should, therefore, be readily digestible. Lunsford & Dunstan (1958) were successful in digesting waste from a pea-blanching factory. The factory produced 819 gallons ($3 \cdot 72 \, \text{m}^3$) waste-water per minute, but $68 \cdot 6 \%$ of the BOD was contained in only 7 % of the water, that from the spray reels of the blanchers and the blanchers themselves. This water contained $3 \cdot 96 \%$ TS of which $92 \cdot 5 \%$ was organic and mostly in solution. It appeared, then, to be reasonable to separate the waste streams and treat the blancher and reel waters by themselves. The digesters used in the tests were single-stage, stirred-tank digesters run at both mesophilic and thermophilic temperatures. Mesophilic and thermophilic floras were developed by adding pea waste, at the appropriate temperature, to domestic sewage digester sludge.

Mesophilic digestion at 37 °C was optimum at a six-day detention time with 85 % of the organic matter removed and a gas production of $12 \cdot 4 \, \text{ft}^3/\text{lb}$ organic matter digester ($0 \cdot 77 \, \text{m}^3/\text{kg}$). The shortest operable detention time was $3 \cdot 5$ days, but here only $11 \cdot 3 \, \text{ft}^3$ of gas was produced per lb of organic matter digested ($0 \cdot 70 \, \text{m}^3/\text{kg}$). The fact that the organic matter was in solution suggests, as previously described, that it would be easily digested and that the limiting reaction would tend to be methanogenesis. This limiting detention time is similar to that found for methanogenesis in piggery-waste digestions, and in pure culture.

Thermophilic digestion at 55 °C could be run at shorter detention times than the mesophilic process. Here, the optimum detention time was 3·5 days when 14 ft³ gas/lb organic matter digested (0·87 m³/kg) was produced. The minimum detention time was two days, when only 11 ft³ (0·69 m³/kg) gas was produced and solids removal was incomplete.

Under optimum conditions the gas produced was similar to that of domestic sewage, being 63·5% methane, 32% carbon dioxide, 3% hydrogen and 1% nitrogen. At maximum loading, gas composition had deteriorated to 35% methane, 63% carbon dioxide and 2% hydrogen.

As pea-processing waste leaves the machines at near boiling temperatures no additional heating should be required for digestion and thermophilic digestion becomes a possibility. In this case calculations showed that a 290 000 gallon (1318 m³) digester would be required if the factory operated for 24 hours per day and some 3·5 million ft³ (9·9 × 10⁴ m³) of gas would be produced. Power inputs would be more than covered by the gas produced.

The use of a fixed-film reactor for the anaerobic digestion of bean-blancher wastes has recently been reported by Van den Berg & Lentz (1979). Preliminary studies showed that maximum loading rates equivalent to 15 kg VS/m³/day could be achieved for a 1·6 cm diameter reactor column. Removal of COD was 80–93% depending on column diameter and direction of flow and daily methane production reached the equivalent of over 5 m³/m³ digester volume.

Knol *et al.* (1978) carried out 90-day experiments on wastes from processing of apples, asparagus, carrots, green peas, French beans, spinach and strawberries, in 1-litre stirred, single-stage digesters at various loading rates at a detention time of 32 days. Feed was added and digester contents removed once a day for four days a week at 40 ml/day and for three days at 20 ml/day. Average biogas yields were equivalent to from 0·30 to 0·58 m³/kg VS added per day. The methane content of gases varied from 64 to 82%. Some of the wastes required nitrogen addition, or alkali addition to overcome low pH caused by rapid fermentation of the sugars of the waste. Digested sewage sludge (1 litre) was used as initial digester inoculum.

Other food-processing wastes have been looked at as potential substrates for anaerobic digestion. For example, Pohland & Hudson (1976) studied aerobic and anaerobic treatments as alternatives for shellfish-processing waste-waters. Using 20-litre continuous-flow, stirred-tank digesters at 37 °C they found that there were some limitations to digestion for waste-water stabilisation. A substrate-associated partial inhibition of methanogenesis was reduced by recycling 50 % of the digester solids and this increased the process efficiency. The methane content of the gas was 70–80%. They concluded that because of the high methane content of the gas, even when some inhibition occurred, there was a definite potential for significant methane production and that further work with recycling and more sophisticated techniques would further justify the process.

Sugar cane and bagasse waste waters have also attracted some interest. Candelario (1974) (reported by Ghosh, 1972) investigated the feasibility of purifying sugar-cane waste-waters by various treatment methods including anaerobic digestion. Experiments were carried out in a 5-gallon (22·7 litre), gas-mixed digester, maintained at 25 °C and operated at detention times between two hours and seven days. Maximum loading was equivalent to 0·64 kg VS/m^3/day. Gas production increased with detention time reaching a peak at four days and then declined as the detention time was further increased to seven days. Gas produced contained 60 % methane and 40 % carbon dioxide, and the pollutional load of the waste was reduced by up to 50 %.

A comparison of the gas production from fresh and stored cattle-waste with that from bagasse and sugar cane has been carried out in small-scale batch digestion (T. R. Preston, personal communication). The digesters were run at ambient temperature which varied from 29° to 37°C. Three percent of urea was added to the sugar cane and bagasse to increase the nitrogen content of the feedstock. The workers found that gas production was significantly greater from sugar cane and bagasse, 3·5 and 3·24 litres respectively per kg dry matter per day being produced as compared to 2·96 litres from the cattle wastes. There was, however, a delay of seven days in gas production from the sugar cane digestion, due to low pH, which was immediately overcome when sodium hydroxide was added on the eighth day.

The upflow, anaerobic, sludge-blanket digester has been extensively investigated in the Netherlands since 1971 (Lettinga *et al.*, 1980) as a method of waste treatment. Using 1- to 60-litre laboratory digesters and later 6 m^3 and 30 m^3 pilot-plant, the process has been applied to industrial wastes including sugar-beet- and potato-processing waters.

Start-up of the process was achieved by using a digested sewage sludge as seed and a low daily waste input of 0·1–0·2 kg COD/kg sludge TS. This was gradually increased until in eight to ten weeks, inputs of 0·8 kg COD/kg solids could be tolerated. The maximum loading rate that could be achieved in laboratory-scale digesters was 10–12 kg COD/m^3/day, except when a particular seed sludge adapted to sugar-beet waste was used in place of sewage sludge. In 6-m^3 pilot-plant digesters higher loading rates were obtained. Digesters were able to stand periods of shut-down of up to a year or more (useful in the treatment of seasonal wastes) and restarted easily within one to three days. Once a suitable sludge blanket has been established for a particular waste it was used to seed future digesters. Digestion appeared to be effective at sub-optimal temperatures when the loading rate was correspondingly reduced and the system was able to function between 10 °C and 45 °C. Gas production in the pilot-scale system could disrupt the sludge blanket and cause problems but problems would be less on a large scale. Depending on the waste to be treated and the conditions applied, an effective treatment could be obtained at high loadings. For example, more than 30 kg COD/m^3/day at 30 °C for potato and sugar-beet waste-waters, at a liquid detention time of as low as three hours was effective.

The fermentation industry is another possible source of feedstock for anaerobic digestion. The industry produces alcohol and alcoholic beverages and the wastes are produced in large quantity and contain protein, cellulose, hemicellulose and lipids along with soluble carbohydrates and alcohol residues. The wastes are amenable to anaerobic digestion and since the fermentation and distillation plant requires large amounts of heat energy there is a use for methane as a fuel on site.

Stander & Snyders (1950) showed that yeast waste liquor, wine pot-still liquor and molasses slop could be digested with a good decrease in pollutional properties of the wastes. Laboratory digesters were semicontinuously inoculated with portions of continuous seed cultures, developed from sewage-digester sludge and the appropriate waste, to improve detention times. (On a large scale a similar effect would be obtained by a feedback process.) From the results they calculated that both wine pot-still liquor and molasses slop could provide sufficient gas to run the digester and to help in evaporating digester effluent for potash recovery. Yeast waste, however, could produce only enough gas to heat the digester.

Rudolphs & Trubnik, as early as 1949, made a study of compressed-yeast-waste treatment and summarised their results in a comprehensive set of papers (Rudolphs & Trubnik, 1949, *a, b, c, d*). They found that in comparison with various physical and chemical methods of treatment anaerobic digestion followed by aerobic trickling filter treatment resulted in a degree of purification high enough to warrant consideration for large-scale operation. Various amounts of spent nutrient medium from the yeast production were added daily to laboratory-scale digesters run at 30 °C and an equal amount of supernatant digested material was decanted. Sludge built up in the vessels. The digesters were either just stood in the laboratory or placed on a shaker platform and shaken for one hour after each feed addition and ten minutes per hour between feeds. Agitation increased gas production and BOD reduction. The sewage-sludge seed had to be acclimatised to the yeast waste. Gas production was 474 litres/kg VS added and TS were reduced by 43 to 60 %. The gas was about 73 % methane, 25 % carbon dioxide and 1·5 % hydrogen sulphide. The solids in the waste were almost entirely colloidal or in solution and would thus be easily available to the digester bacteria; only 100–200 mg of the 10 000–12 000 mg/litre of TS in the waste were Suspended Solids. The maximum loadings appeared to be about 0·04 lb BOD/ft^3 digester capacity for unstirred and 0·13 lb for stirred digestion. These loadings were said to represent minimum theoretical detention times of four days and one day. However, as sludge was building up in the digesters it would appear that the system was to some extent acting as a feedback digester so that bacterial detention times would be greater than liquid detention times.

The authors suggested that on a large scale, stirring could be by mechanical agitation or by introducing the feed into the bottom of the digester by means of a rotating distributor, thus creating a continuous upflow. They also concluded that a continuous or semi-continuous loading of the digester would improve matters still further than the once-daily loading employed. Addition of lime to the digesters was found to have no beneficial effects.

These laboratory experiments served as a basis for pilot-plant experiments for design of a full-scale treatment plant, but as the pilot-plant was as large as some digesters now being built these experiments are discussed in the section dealing with large-scale plant.

Lovan & Foree (1972) looked at the stabilisation of brewery press liquor by anaerobic filter. Since the waste was warm on leaving the factory relatively little heat would be required to maintain mesophilic conditions and even running at thermophilic temperatures might be economically feasible if the gas produced was used to heat the system. With a liquid detention time of 2·15 days and a loading of 100 lb COD/1000 ft^3/day (45·3 kg COD/28·3 m^3/day) there was a COD removal efficiency of 90 % in the bottom 6 in (15 cm) of the filter. Calculations from this suggested an effective loading of six times the above could be obtained.

Schroeder (1975) was reported by Ghosh (1972) to have run laboratory-scale and pilot-plant studies on both aerobic and anaerobic treatment of wine stillage over a two-year period. Experiments were conducted at 57 °C in 1·5-litre continuously-mixed, conventional digesters loaded (with equal removal) once daily. At a 15-day detention time little gas was produced, at a 30-day detention time gas production was 550 ml/day and at 60 day-detention time gas production fell to about 250 ml/day. The pilot-plant studies, were, however, carried out in a kind of anaerobic filter or sludge-retention digester consisting of a 0·92 m diameter × 2·44 m deep steel tank packed with 5 cm Douglas-fir bark chips. Lack of temperature control was said to be a primary factor in the poor results obtained, but three-day liquid detention times were used.

Bhaskeran (1964) experimented with methods of utilising wastes from molasses dilstilleries producing alcohol in India. There were about thirty of these in a small area and they discharged wastes with a polluting power equivalent to the waste produced by ten million people. This was creating serious problems on land and in rivers. Sedimentation and chemical treatments were not effective and aerobic treatments not practicable.

Pilot-plant experiments were carried out in conventional mesophilic, stirred-tank digesters taking 1000 gallons (4·5 m^3) per day at an optimum detention time of 10 days. Gas stirring gave slightly better results than mechanical stirring. Twenty-five volumes of gas (60% methane) were obtained from one volume of spent-wash feedstock. The spent-wash was hot and this heated the digester. Calculations showed that on a full scale the gas produced could supply the entire fuel requirements of the distilleries. Over 80% of the BOD was removed by digestion and a final polishing of the effluent by an activated-sludge unit made it suitable for land or river disposal. By-products from the digestion were vitamin B$_{12}$ and potassium salts.

Basu & Le Clerc (1975) compared the anaerobic digestion of molasses distillery wastes at mesophilic (35°C) and thermophilic (55°C) tempera-tures. Batch and continuous digestions in converted aspirator bottles were used with a digested sewage sludge inoculum. Batch digestions gave good BOD reductions, but only after 100 days digestion. Continuous digesters at 35°C and 55°C and 10 day detention times gave gas productions from 4·5 to 6·75 litres/day. BOD and COD reductions were the main interest, and they concluded that although thermophilic digestion gave slightly better results, mesophilic digestion was more of a practical proposition because of the costs of running a digester at 55°C.

Contact digesters were investigated by Roth & Lentz (1977) as a method for treating rum stillage, but in these experiments reductions in BOD and COD were also the principal interest. The stillage did digest, after addition of small amounts of ammonium and phosphate salts. Shock loadings, and no loading for up to 30 days, did not affect the digestions.

Tofflemire (1972) surveyed methods of treating wine and grape waste-waters and found that digestion had been shown to be practicable in laboratory experiments but was not widely used in full-scale treatment units. He concluded that with proper waste equalisation, close operational control and a high temperature, digestion would probably achieve substantial BOD reductions, although further polishing would be required to complete the treatment.

Other Industries

Factories producing wood pulp, paper and fibre board also produce highly-polluting waste-waters which have been shown by laboratory and pilot-plant experiments to be amenable to digestion. This can be used as a means of removing a large proportion of the BOD from the waste-waters. However, since the degradable material is largely carbohydrate, supplementary nitrogen and phosphorus is needed for many wastes.

Berger in 1958 summarised a number of studies by various workers on pulp- and paper-waste digestion. Neutral sulphide liquors appeared to digest well initially in laboratory tests, but digestion fell off after a time. This was thought to be due to sulphide derived from sulphites and sulphates in the waste. Gas production in all experiments was only about 5 % of that from sewage per kilogram of waste digested. It was thought that if the sulphide toxicity could be overcome digestion would help to purify the waste, but it could not appear to give a worthwhile gas production.

Johnson & Bloodgood (1958) carried out extensive laboratory and pilot-plant experiments on digestion of strawboard-pulping wastes. Digestion was possible at 37 °C and digester loadings equivalent to those of sewage sludge could be used. However, a 19 500 US gallon (7·4 m³) digester designed to treat waste stronger than that in the laboratory experiments gave rather disappointing results.

Satisfactory digestion of paper board white water, supplemented with nitrogen, at 30 °C and a four-day detention time, and rope, rag and jute wastes at six, four and two day detention times (with supplementary nitrogen and phosphorus) were reported in 1950 and 1952 (Tech. Bulletin).

Pharmaceutical-industry wastes have also been looked at.

Heukelekian in 1949 showed that waste penicillin-production broths would digest. Gas production was two to two and a half times the volume of waste feed, and this would be useful on a factory scale. Anaerobic filters were used by Jennett & Dennis (1975) to successfully reduce the polluting power of pharmaceutical waste-waters, although no gas production was given. The filters were of 14·25 litre volume filled with quartzite granules and run at 35 °C. They operated for six months without the need for sludge disposal and shock loadings were easily tolerated.

Anaerobic digestion could also be of use in petrochemical refineries where it has been found that after prolonged acclimatisation to aldehydes, acids, alcohols and esters, these compounds could be broken down with resulting methane production (Chou, 1978). Hydroxyl groups and an increasing carbon chain reduced the toxicity of compounds to the digester flora. Acclimatisation to aromatic ring and double-bond compounds was also possible. Chou concluded that the digestion of petrochemical wastes would not only result in a saving of energy over aerobic processes but would also produce methane on a scale for use as a fuel.

Conclusions

This survey has shown that digestion of many factory wastes is possible. The main object of experiments has often been to find a method of reducing pollution from the factories rather than to produce biogas. Since many of the waste-waters are relatively dilute and in large volume, gas production per unit volume may not be high and conventional single-stage digesters not of use. However, use of feedback digesters or anaerobic filters can give plant of reasonable size and gas in reasonable volume compared with digester size. The gas may be sufficient only for digester heating, but the low energy requirements of digestion and such use of the gas, may make even this a worthwhile process in view of the present energy-saving requirements. But since some factory wastes are hot, digester heating may not be required and if all or even a low specific gas production is available for use then the digestion may be worthwhile from the point of view of reducing factory requirements for external energy.

Even where experiments were successful they were often not carried on to full-scale plant because of the conditions prevailing at the time. The authors are aware, for instance, of one or two cases (not reported here) where factory wastes were amenable to successful digestion, but the process was not proceeded with when permission to pipe wastes to sea was granted. Present and future interest in pollution control and waste utilisation may make schemes for digestion of factory wastes more attractive.

AGRICULTURAL WASTES

Wastes from agriculture, and especially animal excreta from the intensive farm unit or feedlot, form one of the largest sources of biomass in many countries. Some figures and projected figures are given in Appendix 2. The intensive-animal-production excreta are particularly useful for processing, in that they are produced in a confined area and so transport to the digester is relatively easy. On the medium scale there is a local use for the energy in that the intensive farm units are mechanised and with animals housed indoors there is a need for heating or ventilating the animal houses. On the large and very large scale there are possibilities for use of energy in the farm neighbourhood or as part of a national scheme.

Although gas generation on a small scale is possible in some countries, it is the possibility of relatively large-scale energy production from the intensive farm unit which has in the last few years caused an increased interest in anaerobic digestion of farm animal excreta.

The production of energy by digestion of crop residues or energy crops is somewhat more problematical than production of energy from intensive-farm wastes. This is because the crop residues are often left in the fields and there is an energetic and monetary cost in gathering them in for digestion, and similar costs apply to gathering of energy crops, and a further factor involved is that of crop yield and hence gas production per unit of ground area.

Nevertheless, whatever factors have to be considered later, a primary fact to be obtained is that of yield of gas, and optimum conditions for gas production, from a possible feedstock. This information can be obtained from properly conducted small-scale experiments. A large number of experiments have now been done on agricultural wastes, and a number are summarised here.

Piggery Waste

When the authors started small-scale experiments on anaerobic digestion of farm wastes in 1967 the principal reason for the experiments was that digestion seemed to be a possible way of controlling pollution from storing and spreading of animal excreta. Piggery waste was chosen for the first experiments because intensive pig-producing units, with pigs housed in dense populations indoors and on slatted floors without bedding, were probably the largest producers of concentrated animal waste in the UK. Some poultry units produced more waste than the big piggeries, but such units were few, and overall the many intensive piggeries producing an obnoxious liquid waste ('slurry') posed the greater pollution problem. Since then there has been the rise of the large dairy unit and more intensive beef cattle rearing, and the pig industry has somewhat declined. Nevertheless, pig units still produce very large amounts of waste and will continue to do so in the future. In other countries pigs are a major source of animal production and units much larger than those in Britain are in operation. With legislation for the stricter control of pollution now beginning to be enacted in Britain and elsewhere the case for anaerobic digestion as a method of reducing the polluting power of piggery wastes still exists and may in some cases be even greater than before, but present and future prices of fossil fuels and possible shortages have turned the production of energy from digestion of piggery and other animal wastes from a by-product of pollution control to a major reason for running digesters, and one that is economically viable.

The results quoted in this section are, then, principally concerned with gas production although many of the papers contain detailed information on reduction of pollutants.

Taiganides (1963) experimented with laboratory-scale digestion of pig wastes. He concluded that pig excreta could be digested satisfactorily at a VS loading of $3 \cdot 21$ kg/m^3 digester capacity/day, with once-daily loading of the stirred-tank digesters and a detention time of 10 days. The excreta were diluted with tap- or wash-water before digestion as he believed that copper was present in the waste at inhibitory concentrations. Digestion was considered feasible with a feedstock slurry of not more than $3 \cdot 3 \%$ TS concentration.

Schmid & Lipper (1969) reported digestion tests with 5-litre plastic units. The digesters were continuously mixed and maintained at 35 °C initially and later at 20 °C. The units were inoculated with digested domestic sewage and then loaded with undiluted piggery waste at 7–8 % TS. Detention times were 20 days and 10 days. Results were disappointing in that gas production and methane content of the gas decreased with time and volatile acid concentrations increased, indicating digester failure. The authors hypothesised that for piggery waste, if urine was collected along with faeces and no water was added, then conventional digestion could not be practised due to ammonia toxicity; though, because of the high ammonia concentration, very high volatile acid concentrations could be tolerated without a drop in pH. Therefore, the acid-forming bacteria could continue functioning, resulting in the liquefaction of the waste but not stabilisation.

However, few of these experiments were made over a long period of time and loading rates were erratic and changed rapidly. About the same time as the last-mentioned experiments (1967–68) a series of studies on piggery-waste digestion was started in Aberdeen using the 15-litre, stirred (80 rpm stirrer), single-stage, laboratory digesters previously described. These digesters were, as mentioned before, loaded every day, with removal of an equal volume of digester contents, and run, at 35 °C, for long periods at each loading rate. The experiments showed that a digestion could not be built up from a tank of high-solids, undiluted pig excreta (10 % TS). The digestions went acid and ceased to function; volatile acids rose from 2500 mg/litre to 11 700 mg/litre. Pig excreta (2 vol) diluted with water (1 vol) also went acid. The gas produced was 80–95 % carbon dioxide with some hydrogen and a little methane. The experiments were continued with excreta taken from the under-floor troughs of a piggery housing fattening pigs fed on a dry meal (barley). The pigs were on concrete floors with a slatted dunging area. In the first few years some sawdust was used in the pens, but this was eventually discontinued. The waste TS were about 70 % VS plus salts, cement dust and grit.

However, digestion could be started from a digester filled with water to which piggery waste diluted to 2·5 % TS was added at a rate equivalent to a detention time of 35–40 days. A balanced digestion was established in 7–8 weeks and loading rate was then increased. A digester seeded with digested sewage sludge and similarly loaded, established a balanced digestion rather more quickly. These, and other experiments where digesters went acid, showed how critical an initial low loading rate is to trouble-free establishment of digestion. The seeded and unseeded digesters attained the same performance, and both digesters were run for over 18 months at different loading rates. Performance was satisfactory at 14-days detention time when the feedstock was a slurry diluted to less than 4–4·5 % TS, but performance declined when loading rates were increased. This decline in performance was shown to be not due to copper, volatile acid or ammonia concentrations. Gas production reached a maximum of 13–14 litres/digester (15 litre volume)/day at loadings of 0·18–0·20 lb VS/ft^3/day (2·88–3·20 kg/m^3/day). The gas contained about 60 % methane (Shaw, 1971; Hobson & Shaw, 1971, 1973).

It was thought that digester performance would be better the nearer the loading became to the theoretical continuously-fed culture, and these experiments were followed by experiments with the 100-litre, single-stage, pilot-plant previously described. This was loaded every 5 minutes, with a gravity overflow and continuously stirred. These experiments showed that performance was better as expected. The digestion was started from an inoculum of 100 litres of digested sewage sludge at a loading of 2 % TS at a 32 day detention time. Digestion stabilised after about six weeks and loading rate was then increased. An optimum detention time at 35 °C of 10 days was obtained, with gas of 68–70 % methane content, produced at a rate of 0·30 m^3/kg TS added, over a range of loading rates in terms of slurry solids. This gas production rate is equivalent to about $0.43 \, m^3/kg \, VS$ added. The digester pH has always remained about 7.2 and there has been no need to adjust the pH by additions of alkali or acid. The digester has been running continuously from 1971 to the time of writing (without reinoculation), except for some periods of mains electricity failure and a stoppage of six weeks due to a nation-wide outbreak of swine vesicular disease. These breaks showed that the digestion could be easily restarted by a few days of low loading (after reheating when necessary) after periods of no loading with or without heating. Other tests showed that shock loading could be tolerated. Reduction of detention time to seven days caused some fall off in performance, while three-days detention time was near the point of complete breakdown, as previously described.

Gas production per kilogram TS added increased linearly from $0.26 \, m^3$ to $0.42 \, m^3$ when the digester was run at temperatures between $25 \, °C$ and $44 \, °C$. Above $44 \, °C$ digestion rapidly falls off as the limit of mesophilic bacterial growth is reached, and production would be expected to fall off rapidly below $25 \, °C$. Van Velsen (1977) noted that when the temperature control failed on a 240 litre pilot-plant digester running at $32 \, °C$ and the temperature rose to $50 \, °C$ (i.e. that of the hot-water heating coil) for about one day digestion failed and the process took five weeks to fully recover after an initial period of low loading. He also found (1979) that gas production fell off rapidly below $25 \, °C$ and was almost zero at $15 \, °C$. However, for these latter experiments the temperature was changed slowly. A sudden drop in temperature (say $5 \, °C$ in a few hours) resulted in a drastic fall in gas production and some days (seven or eight) of constant temperature and loading were required to recover the previous steady-state in the Aberdeen experiments.

With a feedstock below about $2 \% \, TS$, digester washout can occur. Gas production per kilogram TS added remained virtually constant from 2% to about 5.5–$6 \% \, TS$, but specific gas production then declined until with $9.5 \% \, TS$ feedstock only $0.19 \, m^3/kg \, TS$ was being produced. Total Solids digestion was also reduced. The digestions were, however, stable over long periods at these higher solids loadings.

At a 10-day detention time and over, and the optimum range of solids loading, reductions in VFA, BOD and COD of greater than 90, 80 and 50% were obtained, while ammonia concentrations were virtually those of the feedstock. The digested sludge was stable and odour was very much diminished.

These results showed that digestion of piggery waste could be obtained over a range of loadings and that digestion was stable and resulted in good gas production and reduction in pollutants in the waste (Bousfield *et al.*, 1972, 1973, 1974; Robertson *et al.*, 1975; Summers & Bousfield, 1976; Hobson *et al.*, 1979, 1980; Summers & Bousfield, 1980).

Although digestion efficiency is reduced with a feedstock above 6 % TS, this is probably of little consequence in practice. Neat pig faeces and urine is of about 10 % TS consistency, but the authors have found that in piggeries, water spilled from drinkers and from other sources generally dilutes the slurry, so that slurries of more than 6 or 7 % TS are seldom obtained, and with some drinking systems, or with wash water or rain water getting into the slurry channels the slurry from many piggeries is of about 4 % TS or even less. In addition, slurries of more than about 7 % TS are difficult to pump and stir.

The Total Solids digestion at 35 °C and a 10-day detention time was about 36 %. As VS were about 70 % of the TS this corresponds to a VS digestion of about 50 %. Analysis of all the results showed that at longer detention times only a slight increase in solids digestion took place and that the digestion at infinite detention time would probably be about 42 % of the TS. However, in these experiments the slurry solids were of similar composition, the thin slurries being obtained by dilution of thick slurry. It is possible to obtain slurries of nominally the same solids concentration but in which the solids compositions differ; i.e., by either removing large particles from, or diluting, a thick slurry. The digestion properties of, and solids reduction in, these two feedstocks will differ considerably.

The ammonia concentrations in the feedstocks used in the above experiments were not more than about 2000 mg/litre and ammonia toxicity was not apparent. Higher ammonia concentrations were sometimes found but no inhibition of digestion was noted. Lapp *et al.* (1975) studying piggery-waste digestion in a pilot-plant found that digestion could be carried on successfully with ammonia concentrations above 3000 mg/litre providing the bacteria were acclimatised to these high concentrations. Further work by Kroeker *et al.* (1976) attributed the stability of their piggery-waste digestions to the relatively high ammonia concentrations in the feedstock.

Schmid *et al.* (1975) also found that piggery-waste digesters could have an ammonia concentration exceeding 3000 mg/litre, and they investigated methods for the removal of ammonia. Significant ammonia stripping could be achieved by the removal of carbon dioxide from the digester gas and recirculating the carbon-dioxide-free gas for mixing. This process increased the digester pH to about 8 and had the added advantage of producing high-methane gas. However, if the digester pH was held at 8, significant amounts of ammonia could be stripped by recirculating digester gas from which only the ammonia had been removed.

Van Velsen (1977) in detailed laboratory and pilot-plant studies of mesophilic piggery-waste digestion in stirred, single-stage digesters, found that even with ammonia concentrations up to 4000 mg/litre at pH 7, a stable digestion could be achieved and maintained with feedstocks of 5–11 % TS, and detention times of 10–40 days. However, with the 11 % TS waste at a 10-day detention time gas production was very low, and even at 15–20-day detention digestion was not as good as with lower solids content feedstock. For feedstocks with solids contents around 5–7 %, digestion was better at detention times longer than 10 days, but levelled off at 20 days. This is unlike the authors' experiments with the 100-litre digester where there was little difference between 10 and 20 days. The piggery waste used by Van Velsen also gave rather a lower gas production than the waste used by the present authors. A maximum of about $0 \cdot 169 \, m^3$ methane (equivalent to $0 \cdot 245 \, m^3$ of gas of 69 % methane) was produced per kilogram TS added to the digesters. These results are rather like those of the previously described 15-litre digesters which were fed once daily. Van Velsen's laboratory-scale digesters were fed once per day for five days and then with a double amount at one time during the next two days. The pilot-plant was fed once daily. Thermophilic digestion of piggery waste at 55 °C was greatly influenced by ammonia concentration (Van Velsen, 1979). Although methane was still produced at ammonia nitrogen concentrations of 1500 mg/litre this was at 25 % less than in a parallel digestion at 30 °C. If ammonia nitrogen were increased to 3500 mg/litre, digestion almost ceased and did not recover within a 50 day period.

Fischer *et al.* (1975) monitored a pilot-plant digester ($58 \cdot 4 \, cm$ diameter $\times \, 139 \cdot 7 \, cm$ long) maintained at 35 °C. Stable digestion was obtained at loading rates from 2·38 to 2·86 kg VS/m^3/day (cf. 15-litre digester results previously mentioned). Gas yield was about $0 \cdot 99 \, m^3$/kg VS digested. Digester pH was 7·3 and ammonia nitrogen 1000 mg/litre. Waste from pigs injected with the antibiotics tylosin and lyncomycin disrupted the digestion process.

The use of a two-phase process for piggery waste was investigated by Smith *et al.* (1977). In theory the system consisted of an acid digester, the output from which was fed into an anaerobic filter, both maintained at 35 °C. In practice, however, attempts to establish an acidification phase failed and an acidic substrate was provided for the filter in the form of acid waste-water from beneath the slatted floors of a pig-finishing unit. Once equilibrium had been reached there was mass balance between the total gas production and decrease in volatile acids between input and output. Optimum results were obtained at a detention time of approximately 32 days and a substrate feed rate of 30·8 kg/m³/day. Volatile acids were reduced from 68 000 mg/litre to 2200 mg/litre and 2·54 volumes of gas were produced per day per digester volume.

Newell *et al.* (1979) concluded that the anaerobic filter could be used with strong agricultural wastes, but limitation was imposed by the solids content of the wastes. Wastes with low solids could be fed direct, but high-solids wastes required prior liquefaction and settlement. This latter was fairly easily achieved for piggery-waste and laboratory-scale filters operating at 30 °C, and a detention time of three days achieved an average COD removal of 90 % from the liquefied waste, with a gas production of 0·43 m³/kg COD removed. The gas contained 80–85 % methane.

Ways of increasing the biogasification of pig faeces using a sodium hydroxide treatment were investigated by Ngian *et al.* (1977). Equal weights of sodium hydroxide solution and dry faeces were mixed and left for 14 days and then used as a substrate for a batch laboratory digester (previously seeded with rumen contents) at a loading of 12 g dry matter/litre. The digester was run at 39 °C for 140 hours without further addition of substrate.

Concentrations of up to 7 % sodium hydroxide increased gas production progressively to a limit of 33 % over untreated faeces. Nine percent hydroxide was no different, but 12 % caused a decrease in gas production. The gas contained only 35 % methane, attributed to the rumen contents inoculum as later studies using a digested-sludge inoculum gave gas with more methane.

Although a proper digestion was not established it is evident from these experiments that alkali treatment will enhance biodegradability of faeces, as expected.

Van Velsen (1979) also investigated pretreatment of piggery waste. Studies indicated that decreasing particle size of the waste increased gas production rate but did not affect the ultimate degree of liquefaction. Liquefaction was increased by acid or alkali treatment for one hour at pH 1 or 13. Alkaline treatment was most effective, increasing the degree of liquefaction by 30 %.

Pretreatment for one hour at 100 °C under atmospheric pressure increased liquefaction by up to 80 %. The most effective treatment was one hour at pH 13 and 100 °C, which almost doubled the degree of liquefaction.

Poultry Waste

Poultry excreta is a solid mixture of faeces and urine, of about 70 % water content. If the excreta are from caged birds the material tends to dry on the collecting trays and can be of lower moisture content. The excreta are usually collected and disposed of as a wet solid and so for digestion, water would have to be added to make a slurry for use in the conventional types of digester. However, some poultry farms use water to flush out the excreta from under cages and channel it for disposal, so here the installation of a digester system would not necessarily mean any additional water being added.

In deep-litter systems the excreta drop into straw, shavings, corn-cobs or other fibrous material and the mixture is scraped from the poultry houses for disposal. This type of material, slurried with water, could pose mechanical problems as digester feedstock, because of the large and fibrous particles which could clog pumps and pipes and cause scum problems. Some form of maceration of the feed slurry would probably be needed. It seems likely from the results to be discussed below, that the high ammonia content of poultry excreta would prevent, or make difficult, a successful solid-state digestion (see later).

Most experiments on poultry-waste digestion have been carried out with the faeces–urine mixture of 'pure' poultry droppings.

Baines (1970) found in small-scale experiments with poultry excreta that gas production was a maximum of $0.5 \, ft^3$ $(0.014 \, m^3)$/hen/day. Each hen produced, on average, $0.33 \, lb$ $(0.15 \, kg)$ of excreta of 25% solids. To this was added $0.6 \, lb$ $(0.27 \, kg)$ of water to give a slurry of 8% TS for digestion.

Gramms *et al.* (1971) found, in laboratory digesters, an organic matter digestion of $57–68 \%$ of the poultry excreta, and about $7.3–8.6 \, ft^3$ gas/lb organic matter digested $(0.45–0.54 \, m^3/kg)$.

Long-term tests in the authors' laboratories using the 150-litre, continuously-stirred, single-stage digesters, showed that excreta scraped from under the cages of laying hens, and containing grit from the feed, broken eggs, feathers and other debris, could be digested (Summers & Bousfield, 1978; Hobson, 1979*a*; Hobson *et al.*, 1979, 1980; Bousfield *et al.*, 1979). The excreta were slurried with water to the required solids content. Gas production (70% methane) was (from a 6% TS slurry) $0.362 \, m^3/kg$ TS added at a 15-day detention time and $0.380 \, m^3/kg$ at a 20 day detention at $35 \, °C$. These and other experiments suggested an 18–20-day detention time for good digestion. Breakdown of Total Solids was only about 23% at the 20 day detention and as the volatile acids content of the slurry is high ($10\,000$ mg/litre or more in a 6% TS slurry) a large proportion (some 25% depending on the slurry) of the gas comes from breakdown of these acids.

The ammonia content of the waste was high and increased with increasing solids content of the slurry (from about 2800 mg/litre in a 4% slurry to about 6000 mg/litre in a 13% slurry). Ammonia inhibition probably accounts for decreasing gas production with increasing solids, a 4% TS slurry giving 0·48 m³/kg TS added and a 12% TS slurry giving 0·29 m³ gas. Volatile Solids were about 72% of Total Solids. In practice, an optimum slurry concentration would have to be found by balancing total gas production from the digester at different solids concentrations with the digester size and the amount of water required, and problems of pumping and stirring the slurries.

Hassan *et al.* (1975 *a*, *b*) concluded from laboratory and pilot-plant experiments that 35 °C was the optimum digester temperature and gas yield was better in the mesophilic than the thermophilic range. Solids concentrations of between 7 and 7·5% were optimum both for gas production and organic solids destruction. Trials with a 2·63 m³ digester running at 35 °C and 7–7·5% solids input gave yields of 0·133–0·161 m³ gas/kg solids input.

Wong Chong (1975) examined the digestion of poultry and cattle wastes under relatively dry conditions in batch digesters. A 3 ft³ digester was loaded with dairy manure at 20·8% TS concentration, and inoculated with 5% of sludge from a previous digestion. The digester was run at ambient temperature, the temperature of the contents fluctating from 70 °F to 82 °F (21 °C to 28 °C). There was a linear increase in gas production for 20 days, then a constant production for 40 days and then a decline to almost nothing at about 140 days. Over the 140 days 36·3% of the VS were degraded and gas production was 11·3 ft³/lb VS destroyed overall and the gas contained 60–65% methane. The pH remained at 7·4. All attempts to similarly digest poultry manure failed because of the high ammonia concentrations.

Cattle Waste

Cattle waste can consist of faeces and urine, which makes a slurry of about 12–14% TS (too thick for digester operation); the faeces and urine diluted with washing or other water, or the faeces and urine mixed with fibrous bedding of various kinds from sawdust to long fibres. The excreta vary somewhat in composition depending on the feed of the animals and also on the type of animal. Excreta from milking cows can be different from those of fattening beef cattle and young stock.

Since cattle feed has passed through a microbial fermentation process in the rumen and a further microbial action in the caecum, the faecal residues contain little material easily degradable by the digester bacteria and the gas production would be expected to be lower than that from pigs and poultry which do not have the rumen fermentation.

Hart (1963) found faeces and urine from dairy cows to be only slightly digestible, some 10–16% of the organic matter being degraded. Gas production was 15 ft³/lb organic matter degraded (0·93 m³/kg), a figure similar to that for domestic sewage, but, overall, gas production per kilogram organic matter added to the digester was only about 25% of that from domestic sewage.

Gramms *et al.* (1971) got rather better organic matter digestion (18–26%), but less gas per kilogram organic matter digested (0·16 m³) in laboratory tests on dairy-bull wastes.

The authors found that digestion of mixed excreta from fattening cattle, fed on a number of concentrate, silage, hay and grass diets, in a 150-litre continuously-stirred digester at 35 °C required a 20-day detention time. Gas production (55–60% methane) was, with a 6% TS input, 0·215 m³/kg TS added to the digester at 20-day detention time and 0·195 m³ at a 10-day detention. At the 20-day detention time digestion and gas production increased with increasing solids concentration in the input slurry. At 4·8% TS input gas production was 0·189 m³/kg TS added, rising to 0·260 m³ at 10% TS input. Gas production was levelling out between 8 and 10% TS. This was ascribed, by analogy with rumen fermentations, to too low an ammonia concentration in the low solids waste (prepared by dilution of the thick slurry collected) for optimum bacterial action. At 5% the slurry ammonia concentration was only 510 mg/litre. The cattle were housed on slatted floors without bedding.

Dairy cattle wastes were less digestible, average gas production (55–60% methane) being 0·206 m³/kg TS added with an input slurry of 7% TS content and a 21-day detention time at 35 °C. Gas production was lower at a 15-day detention time, and also lower at 25 °C than at 35 °C. The cattle were fed on silage plus oats and some distillers residues. These wastes were of about 70% organic solids in the TS and contained no bedding material. The 20-day detention time should be sufficient for digestion of straw or similar bedding added to the excreta (see later). (Hobson *et al.*, 1979, 1980; Hobson, 1979a; Bousfield *et al.*, 1979, unpublished).

Converse *et al.* (1977) compared the digestion of three kinds of dairy cattle manure at mesophilic (35 °C) and thermophilic (60 °C) temperatures. The digesters were of the stirred-tank, single-stage type, of 0·71 m³ volume with gas-stirring and electric heaters. Mixing was intermittent at 3 min/half-hour and 20 minutes prior to the once-daily feeding, and removal of digested sludge.

The three wastes studied were faeces alone, faeces plus urine, and faeces, urine plus straw. The wastes were diluted to give a slurry of approximately 6·5 % VS. Detention times were 15, 10·4 and 6·2 days and loadings were 4·25–10·51 kg VS/m^3 digester volume/day.

Gas productions ranged from 1·26 to 1·66 m^3/m^3 digester volume/day for the mesophilic digester, with the gas containing 52·8 to 56·7 % methane. Gas production from the thermophilic digester was from 1·06 to 2·59 m^3 with 49·2 % to 54·5 % methane. For both digesters, VS and COD reductions fell as the detention time decreased and the mixture containing straw yielded less gas than the faeces or faeces–urine, indicating that the straw did not digest readily. Gas production increased on a volume/digester volume/day basis as the loading rate of the thermophilic digester increased. Mesophilic gas production increased only slightly with loading rate. A minimum of 15-days detention time was recommended for mesophilic digestion of cattle wastes. Net energy output averaged 69 % of the gross energy output in mesophilic digestion at a 15-day detention time and 40·3 % for the thermophilic digestion. In both cases heating digester feedstock was the greatest energy input (cf. discussion in previous section).

Laboratory batch digestion of cattle wastes was tested by T. R. Preston *et al.* (personal communication) in Santo Domingo. Gas production from waste from cattle fed on sugar cane almost doubled when digester temperature was increased from 25 °C to 35 °C. There seemed to be little benefit from mixing the digester for waste from cattle fed on sugar cane, but there were indications that mixing was advantageous when the substrate contained longer fibres (e.g. bagasse). Between 29 °C and 36 °C, digestion of TS was 47·3 % and 23·5 % for detention times of 20 and 10 days, and gas production was 49 litres and 25·6 litres per kg solids added at these detention times.

Varel *et al.* (1977) studied thermophilic digestion of faeces and urine from dairy-beef cattle fed on a concentrate ration containing 89·5 % of maize and oats plus some alfalfa silage. They used 3-litre, once-daily fed digesters shaken on a shaker platform and heated by electric tapes around the bottles. The optimum temperature was 60 °C and raising this to 65 °C drastically lowered efficiency. At VS concentrations in the feed of between 2 and 10 %, VS destruction increased with detention time from three to twelve days, being about 51–53 % at twelve-day detention for 4–10 % VS and about 35 % at three-day detention. Increasing feed VS concentration above about 10 % resulted in decreased and unstable digestion. The maximum VS in the feedstock for stable digestion at three, six, nine, and twelve-day detention times were 8·2, 10·0, 11·6 and 11·6% and complete digestion failure occurred at about 14% VS. The effects of the higher solids loadings were suggested as due to ammonia toxicity. The gas production increased with increasing detention time, and the gas contained 52–57% methane. The gas production is reported as 'methane' and methane production at three-, six-, nine- and twelve-day detention times and optimum VS inputs were 0·16, 0·18, 0·20 and 0·22 m^3/VS added. The VS of the waste were about 78% of the TS. A gas production of 0·22 m^3 methane/kg VS is approximately equivalent to a digester gas (57% methane production) of 0·30 m^3/kg TS.

These results suggest that stable, thermophilic digestion could be run at shorter detention times than mesophilic digestions. However, the three day detention time is probably too short for commercial running unless the feed inputs, etc., are very strictly controlled. To allow a margin of safety, a six- or eight-day detention would probably be more feasible. The feeds of these fattening cattle contained little fibrous material and it could be that longer detention times would be needed for waste from animals fed entirely on silage and hay.

A series of laboratory studies on the feasibility of plug-flow, tubular, digesters as applied to dairy-cow manure was reviewed by Jewell in 1980. Attempts to simulate a plug-flow digester using six-inch diameter tube were not successful, but a series of four, completely mixed, reactors was considered to be an approximation to a tubular reactor (cf. previous section on theory of digesters). Using a 12-day detention time over the series, methane production was 0·71 m^3/m^3 digester/day and a solids destruction of 29% was achieved at 35°C and 8% TS input slurry. Similar results were obtained in a pilot-scale 5·6 m^3 'tubular', plug-flow digester and a larger version is described later.

A 5 m^3, square-shaped, stirred digester was also studied. When operated under similar conditions to the plug-flow digester similar results were obtained. When, however, between-feeding periods were extended to four and seven days, a periodicity in gas production, volatile acids and digester pH appeared. A change from daily to weekly loading at a 30-day detention time reduced gas production by 35%. Addition of straw bedding to digester feedstock (0·93 kg/cow/day) increased gas production by more than 20%.

Animal Excreta Plus Other Wastes

Animal excreta are high in nitrogenous compounds and generally in ammonia content, so that addition of extra carbohydrate could improve digestion. Various vegetable matters, high in carbohydrate, are also available seasonally on farms, and often in temperate climates in winter when extra digester gas production would be useful.

The authors found that grass-silage seepage liquid digested well when added to a piggery-waste digestion at a 10-day detention time and 35 °C. Ten percent by volume of liquid added to a 3·5 % TS pig slurry increased gas production by 16 %. Rotten potatoes also digested easily, and added at 15 % of the piggery-waste TS (3·5 % TS input), increased gas production by 27 %. Later experiments showed that potatoes could be digested with low-solids pig slurry. For instance an 8·5 % TS slurry made up of 2·5 % piggery waste and 6 % potato solids gave 0·500 m³ gas/kg TS added. The gas was lower in methane content (57–58 %) than that from piggery waste alone, as might be expected.

Barley straw did not digest with piggery waste at a 10-day detention time, the time had to be extended to twenty days and some weeks allowed for a suitable cellulolytic bacterial population to build up in the digester before good digestion was obtained. Even so the straw was only 35 % degraded when added at the rate of 1 % TS plus 3·3 % piggery waste TS. Gas production from the straw was about 0·198 m³/kg straw added, and the methane content of the gas was not appreciably changed from that of piggery-waste gas (Bousfield *et al.*, 1974; Summers & Bousfield, 1976; Summers, 1978; Hobson, 1979*b*).

Some other experiments on addition of straw to animal-waste digestion have been previously mentioned.

Hills (1979) more recently carried out laboratory trials on screened dairy-cow manure combined with glucose and cellulose in different ratios. With an increasing ratio of carbon to nitrogen in the feedstock, methane content of the gas decreased from 67 % to 51·7 % (C:N ratios 8:1 and 57·1:1). Optimum methane production occurred at an available C:N ratio of 25:1.

Crops and Crop Residues

Some experiments on combining of crop wastes and animal excreta have been mentioned, but there have been other experiments on crop wastes alone. The subject of crops grown specially for digestion is rather outside the scope of this book, but some mention will be made here, for the reasons given in a later section.

Kanoksing & Lapp (1975) used 4-litre digesters to investigate gas production from oatstraw. At 35 °C gas production was 0·41 m³/kg straw added for batch digesters and 0·38 m³/kg for once-daily fed continuous digesters. The gas contained 52–56 % methane. They concluded that difficulties of handling straw on a large scale would make commercial gas production not economically feasible.

Pfeffer (1980) described thermophilic digestion of wheat straw at 59 °C. At 13·7-day detention time methane production was 0·167 m^3/kg straw fed and as the gas was 58 % methane this is equivalent to about 0·288 m^3 digester gas. Digestion increased with detention time and was about 38 % at 13·7-day detention time. Calculations suggested a maximum of 50 % degradation at infinite detention time. The straw used here was, however, pretreated with sodium hydroxide.

Badger *et al.* (1979) investigated small-scale batch digestions of various crops and crop wastes, at 37 °C and with loading concentrations of 3–10 % TS. Yields of gas decreased with increased solids and above 5 % TS high-soluble-carbohydrate substrates required continuous alkali additions to maintain digester pH at 7. Gas was produced in different amounts and at different rates. Digestion times of 17–36 days were required. The VS destruction seemed higher in these digestions than in the continuous digestions previously noted. For instance, wheat straw (ground) gave 0·53 m^3/kg TS added, with gas of 58 % methane at 5 % TS loading. Grinding the straw increased gas production from the 0·412 m^3 of the chopped straw.

Pfeffer (1980) also tested thermophilic (58 °C) digestion of ground corn-stover. Gas of about 53 % methane was produced and methane production was about 0·125 m^3/kg VS added at a 14-day detention time.

Water hyacinths have been suggested as a crop grown for methane production. Wolverton *et al.* (1975) investigated mesophilic digestion of this vegetable material. As said before, the results are rather outside the scope of the book, but one result is rather surprising and of general interest. Nickel and cadmium contamination of the plants increased gas production from 51·8 ml/day for the non-contaminated vegetation to 81 ml/day. The gas was also said to contain 91 % methane.

Small-scale digestion of marine plants such as kelp has also been investigated as a basis for possible large-scale energy production (e.g. Troiano *et al.*, 1976).

CONCLUSIONS

Small-scale digestions show that most organic materials will digest. Only some representative results have been quoted here, there are other papers on the use of small-scale tests. Some uses of the small-scale digestions in investigating details of digestions already carried out on a large scale have been mentioned. Although some of the experimental techniques used have had little relationship to large-scale practice, others have been such that they could be used to design large-scale plant, and some of the small-scale experiments have been followed by the building of large-scale digesters. The data from small-scale experiments can also be used for model making in various ways. The more simple models can calculate the feasibility of energy production from a particular farm or other digester plant (e.g. Hobson, 1979c). This type of model takes into account only the broad factors of energy production and usage. On the other hand the small-scale data can be combined with data on engineering aspects and costs to get a detailed model of biogas production from different types and sizes of plant. Wentworth *et al.* (1979) and Ashare *et al.* (1979) produced such detailed, technical and economic models for gas production from animal wastes, and there are others. The authors mentioned concluded that gas production from animal wastes was well worth development.

Data can be selected from the small-scale results. For example the data from the mesophilic digestion of piggery waste can be used to calculate an optimum digester running temperature at different ambient temperatures. It may be worthwhile overall to have a lower gas production rate at a lower running temperature but not use so much of the available energy in heating the digester and feedstock. Horton (1980) gives an example of this type of calculation. On the other hand, if an engine is to be run on the gas then calculations may show that for the required power output engine heat is more than sufficient to heat the digester. It might then be worth increasing the digester temperature to obtain extra gas for other uses.

REFERENCES

ASHARE, E., WENTWORTH, R. L. & WISE, D. L. (1979). *Resource Recovery and Conservation* **3**, 359.
BADGER, D. M., BOGUE, M. J. & STEWART, D. J. (1979). *N.Z. J. Sci.* **22**, 11.
BAINES, S. (1970). In: *Proc. Symp. Farm Wastes, Inst. Water Poll. Conf.*, Univ. Newcastle-on-Tyne, p. 132.
BARTH, E. F. (1967). *J. Water Pollution Control Federation* **37**, 86.
BASU, A. K. & LECLERC, E. (1975). *Water Res.* **9**, 103.
BERGER, H. F. (1958). In: *Biological Treatment of Sewage and Industrial Wastes* **2**, 136.
BHASKERAN, T. R. (1964). *Adv. Water Poll. Res.* **2**, 85.
BONOMO, L. (1975a). *Chem. Abstr.* **83**, 209174t.

BONOMO, L. (1975*b*). *Chem. Abstr.* **82**, 174887c.

BOUSFIELD, S., HOBSON, P. N. & SUMMERS, R. (1972). *Proc. Farm Waste Disp. Conf.*, ARC, p, 65.

BOUSFIELD, S., HOBSON, P. N. & SUMMERS, R. (1974). *J. Appl. Bact.* **37**, xi.

BOUSFIELD, S., HOBSON, P. N. & SUMMERS, R. (1979). *Ag. Wastes* **1**, 161.

BOUSFIELD, S., HOBSON, P. N., SUMMERS, R. & ROBERTSON, A. M. (1973). *Farmers Weekly*, (November) p. 79.

BUHR, H. O. & ANDREWS, J. F. (1977). *Water Res.* **11**, 129.

CANDELARIO, R. M. (1974). U.S. Natl. Tech. Inform. Serv., Springfield, Va., W74-12865.

CHEN, K. Y. (1974). *J. Water Pollution Control Federation* **46**, 2263.

CHENG, M. H. (1975). *J. Water Pollution Control Federation* **47**, 362.

CHOU, W. L. (1978). *Diss. Abstr. Int. B.* **38**, p. 5254.

CONVERSE, J. C., GRAVES, R. E. & EVANS, G. W. (1977). *Trans ASAE*, p. 336.

COONEY, L. C. & WISE, D. L. (1975). *Biotech. Bioeng.* **17**, 1119.

DAVIS, J. A. III and JACKNOW, J. (1975). *J. Water Pollution Control Federation* **47**, 2292.

DREW, E. A. & SWANWICK, J. D. (1962). *Publ. Wks. Mun. Serv. Cong.*, p. 1.

FAIR, G. M. & MOORE, E. W. (1932). *Sew. Wks. J.* **4**, 589.

FAIR, G. M. & MOORE, E. W. (1937). *Sew. Wks. J.* **9**, 3.

FISCHER, J. R., SIEVERS, D. M. & FULHAGE, C. D. (1975). In: *Energy, Agriculture and Waste Management* (ed. W. J. Jewell). Ann Arbor Sci. Publishers, Michigan. p. 307.

GHOSH, M. & ZUGGER, P. (1973). *J. Water Pollution Control Federation* **45**, 424.

GHOSH, S. (1972). *J. Water Pollution Control Federation* **44**, 948.

GOLUEKE, C. G. (1958). *Sew. Ind. Wastes* **30**, 1225.

GOSSETT, J. M. & MCCARTY, P. L. (1975). *68th An. Meet. Am. Inst. Chem. Eng.*, Los Angeles.

GOSSETT, J. M., MCCARTY, P. L., WILSON, J. C. & EVANS, D. S. (1978). *J. Water Pollution Contròl Federation* **50**, 533.

GRAMMS, L. C., POLKOWSKI, J. & WITZEL, S. A. (1971). *Trans. Am. Soc. Agric. Eng.*, p. 7.

HART, S. A. (1963). *J. Water Pollution Control Federation* **35**, 748.

HASSAN, A. E., HASSAN, H. M. & SMITH, N. (1975*a*). In: *Energy, Agriculture and Waste Management* (ed. W. J. Jewell). Ann Arbor Sci. Publishers, Michigan, p. 289.

HASSAN, H. M., BELYEA, D. A. & HASSAN, A. E. (1975*b*). In: *Managing Livestock Wastes*. Proc. 3rd Int. Symp. Livestock Wastes, Univ. of Illinois. p. 244.

HAUG, R. T., STUCKEY, D. C., GOSSETT, J. M. & MCCARTY, P. L. (1978). *J. Water Pollution Control Federation* **50**, 73.

HAYES, T. D. & THEIS, T. L. (1978). *J. Water Pollution Control Federation* **50**, 61.

HEMENS, J. & SHURBEN, D. G. (1959). *Food Trade Rev.* **29**, 2.

HEUKELEKIAN, H. (1930). *Sew. Wks. J.* **2**, 219.

HEUKELEKIAN, H. (1949). *Ind. Eng. Chem.* **41**, 1535.

HEUKELEKIAN, H. & KAPLOWSKY, A. J. (1948). *Sew. Wks. J.* **20**, 806.

HILLS, D. J. (1979). *Ag. Wastes* **1**, 267.

HOBSON, P. N. (1965*a*). *J. Gen. Microbiol.* **38**, 161.

HOBSON, P. N. (1965*b*). *J. Gen. Microbiol.* **38**, 167.

220   *Methane Production from Agricultural and Domestic Wastes*

HOBSON, P. N. (1979a). In: *Energy From the Biomass*, Watt Committee, on Energy, London, p. 37.

HOBSON, P. N. (1979b). *Straw Decay, its Effect on Disposal and Utilisation.* Wiley, London, p. 217.

HOBSON, P. N. (1979c). *Processes for Chemicals from some Renewable Raw Materials.* Inst. Chem. Eng., London. p. 1.

HOBSON, P. N., BOUSFIELD, S., SUMMERS, R. & MILLS, P. J. (1979). In: *Engineering Problems with Effluents from Livestock*, EEC Commission, p. 492.

HOBSON, P. N., BOUSFIELD, S., SUMMERS, R. & MILLS, P. J. (1980). In: *Anaerobic Digestions* (eds. D. A. Stafford, B. I. Wheatley and D. E. Hughes) Proc. 1st Int. Symp. An. Dig., Cardiff. Applied Science Publishers, London.

HOBSON, P. N. & SHAW, B. G. (1971). In: *Microbiol Aspects of Pollution* (eds. G. Sykes and F. A. Skinner). Academic Press, London, p. 103.

HOBSON, P. N. & SHAW, B. G. (1973). *Water Res.* **7**, 437.

HOBSON, P. N. & SMITH, W. (1963). *Nature* **200**, 607.

HOBSON, P. N. & SUMMERS, R. (1967). *J. Gen. Microbiol.* **47**, 53.

HORTON, R. (1980). In: *Anaerobic Digestion* (eds. D. A. Stafford, B. I. Wheatley and D. E. Hughes) Proc. 1st Int. Symp. An. Dig., Cardiff. Applied Science Publishers, London.

JENNETT, J. C. & DENNIS, N. D. (1975). *J. Water Pollution Control Federation* **47**, 104.

JEWELL, W. J. (1980). In: *Anaerobic Digestion* (eds. D. A. Stafford, B. I. Wheatley and D. E. Hughes) Proc. 1st Int. Symp. An. Dig., Cardiff. Applied Science Publishers, London.

JOHNSON, C. C. & BLOODGOOD, D. E. (1958). In: *Biological Treatment of Sewage and Industrial Wastes* **2**, 115.

KANOKSING, P. & LAPP, H. M. (1975). Paper presented at Can. Soc. Ag. Eng. Ann. Meeting, Brandon Univ., Brandon, Manitoba, Canada.

KLEIN, S. A. (1972). *Compost Sci.* **1**, 6.

KNOL, W., VAN DER MOST, M. M. & DE WAART, J. (1978). *J. Sci. Fd Agric.* **29**, 822.

KOCH, C. M., NELSON, M. D. & DIMENNA, R. (1978). *Water Waste Engng.* **15**, 27.

KROEKER, E. J., LAPP, H. M., SCHULTE, D. D., HALIBURTON, J. D. & SPARLING, A. B. (1976). Paper presented at Can. Soc. Ag. Eng. Ann. Meeting, Halifax, Nova Scotia.

LAPP, H. M., SCHULTE, D. D., KROEKER, E. J., SPARLING, A. B. & TOPNIK, B. H. (1975). Paper presented at 3rd Int. Symp. Livestock Wastes, Champagne, Urbana, Ill.

LAWRENCE, A. W. & MCCARTY, P. L. (1965). Paper presented at 38th Ann. Conf. WPCF, Atlantic City, New Jersey.

LETTINGA, G., VAN VELSEN, A. F. M., DE ZEEUW, W. & HOBMA, S. W. (1980). In: *Anaerobic Digestion* (eds. D. A. Stafford, B. I. Wheatley and D. E. Hughes) Proc. 1st Int. Symp. An. Dig., Cardiff. Applied Science Publishers, London.

LOVAN, C. R. & FOREE, E. G. (1972). *Brew. Dig.* **47**, 66.

LUNSFORD, J. V. & DUNSTAN, G. H. (1958). In: *Biological Treatment of Sewage and Industrial Wastes* **2**, 107.

MCCARTY, P. L., YOUNG, L. Y., GOSSETT, J. M., STUCKEY, D. C. & HEALEY, J. B. JR. (1976). In: *Microbial Energy Conversion* (eds. H. G. Schleigel and J. Barnea). Verlag and Goltze, Gottingen, p. 179.

MERKENS, J. C. (1962). *Lab. Pract.* **11,** 930.

NEUFELD, R. D. & HERMANN, E. R. (1975). *J. Water Pollution Control Federation* **47,** 210.

NEWELL, P. J., COLLERAN, E. & DUNICAN, L. F. (1979). Poster presentation at 1st Int. Symp. An. Dig., Cardiff.

NGIAN, M. F., NGIAN, K. F., LIN, S. H. & PEARCE, G. R. (1977). *J. Environ. Eng. Div.* **103,** 1131.

PFEFFER, J. T. (1973). In: *Proc. Int. Biomass Energy Conf.*, Biomass Energy Inst. Inc., Winnipeg, Canada.

PFEFFER, J. T. (1980). In: *Anaerobic Digestion* (eds. D. A. Stafford, B. I. Wheatley and D. E. Hughes) Proc. 1st Int. Symp. An. Dig., Cardiff. Applied Science Publishers, London.

PFEFFER, J. T. & KHAN, K. A. (1976). *Biotech. Bioeng.* **18,** 1179.

PFEFFER, J. T. & LIEBMAN, J. C. (1974). Ann. Prog. Rept. to NSF Grant G1 39191.

POHLAND, F. G. & HUDSON, J. W. (1976). *Biotech. Bioeng.* **18,** 1219.

PRETORIUS, W. (1971). *Water Res.* **5,** 681.

ROBERTSON, A. M., BURNETT, G., BOUSFIELD, S., HOBSON, P. N. & SUMMERS, R. (1975). In: *Managing Livestock Wastes.* Proc. 3rd Int. Symp. Livestock Wastes, Univ. of Illinois. p. 544.

ROTH, L. A. & LENTZ, C. P. (1977). *J. Can. Inst. Food Sci. Technol.* **10,** 105.

RUDOLPHS, W. & HEUKELEKIAN, H. (1930). *Ind. Eng. Chem.* **22,** 96.

RUDOLPHS, W. & TRUBNIK, E. H. (1949*a*). *Sew. Wks. J.* **21,** 101.

RUDOLPHS, W. & TRUBNIK, E. H. (1949*b*). *Sew. Wks. J.* **21,** 295.

RUDOLPHS, W. & TRUBNIK, E. H. (1949*c*). *Sew. Wks. J.* **21,** 701.

RUDOLPHS, W. & TRUBNIK, E. H. (1949*d*). *Sew. Wks. J.* **21,** 1029.

RUDOLPHS, W. & ZELLER, P. J. A. (1932). *Sew. Wks. J.* **4,** 711.

SANDERS, D. A., MALINA, J. F. JR., MOORE, B. E., SAGIK, B. P. & SORBER, C. A. (1979). *J. Water Pollution Control Federation* **51,** 333.

SCHMID, L. A. & LIPPER, R. I. (1969). In: *Proc. Animal Waste Man. Conf.*, Cornell Univ., p. 50.

SCHMID, L. A., LIPPER, R. I., KOELLIKER, J. K., CATE, C. A. & DABER, J. W. (1975). In: *Managing Livestock Wastes.* Proc. 3rd Int. Symp. Livestock Wastes, Univ. of Illinois. p. 248.

SCHROEDER, E. D. (1975). Environ. Protective Technol. Ser. EPA-660/2-75-002.

SHAW, B. G. (1971). Ph.D. Thesis, University of Aberdeen.

SMITH, R. E., REED, M. J. & KIKER, J. T. (1977). *Trans ASAE*, p. 1123.

Standard Methods for the Examination of Water and Waste Water (1965). Am. Pub. Health Assoc. N.Y.

STANDER, G. J. & SNYDERS, R. (1950). *Proc. Inst. Sew. Purif.* **447,** 458.

STEFFEN, A. J. (1958). *Biological Treatment of Sewage and Industrial Wastes*, vol. II, Reinhold Pub. Corp., N.Y.

SUMMERS, R. (1978). In: *Rept. on Straw Utilisation Conf.*, MAFF, p. 84.

SUMMERS, R. & BOUSFIELD, S. (1976). *Proc. Biochem.* **11,** 3.

SUMMERS, R. & BOUSFIELD, S. (1978). In: *Proc. Seminar An. Dig.*, ADAS, Reading.

SUMMERS, R. & BOUSFIELD, S. (1980). *Ag. Wastes* **2,** 61.

SWANWICK, J. D., WHITE, K. J. & DAVIDSON, M. F. (1961). *Proc. Symp. Treatment of Ind. and Sew. Sludges*, Inst. Sew. Purif.

TAIGANIDES, E. P. (1963). Ph.D. Thesis, Iowa State Univ. of Sci. and Technol.

Tech. Bull. No. 36 (1950). Nat. Council for Stream Improvement.

Tech. Bull. No. 52 (1952). Nat. Council for Stream Improvement.

TOFFLEMIRE, T. J. (1972). *Am. J. Enol. Viticult* **23**, 165.

TORIEN, D. F. (1967). *Water Res.* **1**, 147.

TROIANO, R. A., WISE, D. L., AUGENSTEIN, D. C., KISPERT, R. G. & COONEY, C. L. (1976). *Resource Recovery Conserv.* **2**, 171.

VAN DEN BERG, L. & LENTZ, C. P. (1979). Poster presentation 1st Int. Symp. An. Dig., Cardiff.

VAN VELSEN, A. F. M. (1977). *Neth. J. Agric. Sci.* **25**, 151.

VAN VELSEN, A. F. M. (1979). In: *Energy Problems with Effluents from Livestock*. EEC Commission, p. 476.

VAREL, V. H., ISAACSON, H. R. & BRYANT, M. P. (1977). *Appl. Environ. Microbiol.* **33**, 298.

WENTWORTH, R. L., ASHARE, E. & WISE, D. L. (1979). *Resource Recov. Conserv.* **3**, 343.

WOLVERTON, B. C., McDONALD, R. C. & GORDIN, J. (1975). Tech. Memo. X-72725, National Aeronautical and Space Agency, USA.

WONG CHONG, G. M. (1975). In: *Energy, Agriculture and Waste Management*, Ann Arbor Sci. Publishers, Michigan. p. 361.

CHAPTER 8

Energy Production by Practical-scale Digesters

The types and amounts of wastes that can be used as digester feedstock are many and varied. As described in Chapter 7, much work has been carried out with pilot-plant studies of digestion of these wastes, but only a little has been done with full commercial-scale plant. The exception is domestic sewage-sludge digestion. Even in this field, where large-scale digestion started about 70 years ago, little detailed information has been published on the running of plants. For the other wastes, large-scale digesters are only just beginning to be built and tested. Because these digesters are in the developmental stage the results of long periods of continuous running are not generally available. Many have been started only a year or less ago and part of this time has been taken up with bringing the digesters up to working loading and making adjustments to running parameters.

In this chapter some data from large-scale plants will be given to show that biogas production and use is possible on a large scale with automated digesters. However, as previously pointed out the large, automated digester is not necessarily required for every situation. The small, hot-climate, digester is a suitable type of digester in its particular context. So, although these digesters may be only the size of some described as 'pilot-plant' in other parts of the book, their running will be considered here, as they are 'practical' digesters.

DOMESTIC AND MUNICIPAL SEWAGE

Mesophilic Digestion
Domestic sewage digestion was developed from the 'cess-pool' and later the 'septic-tank'. Ambient temperature digestion in open tanks followed as a means of stabilising sewage sludges, and this was replaced by the closed-tank digester and finally the heated, stirred, high-rate digester of the present day. In Birmingham (England) tests were started in 1911 on the digestion of all the sewage sludge from a population of one million and the use of gas for running engines which could power electricity generators. Since then,

anaerobic digestion plants have been built in sewage works all over the world, and the gas has been used for heating, lighting and engines.

The construction and running of sewage digesters has been described in a previous section. The mesophilic running temperatures are from about 20 °C to 35 °C.

Grit is removed from the inflowing sewage water. Larger pieces of debris are screened off and these can be macerated and added to the smaller debris which is then settled out from the sewage and forms the 'primary sludge'. The sewage water, a dilute solution and colloidal suspension of waste, passes from the settling tanks to the aerobic treatment plant. This is usually an activated sludge system, although trickling filters are still used in some plants. Growth of micro-organisms in the activated sludge plant, followed by settling-out of the micro-organisms and entrapped sewage particles, reduces the polluting properties of the water to a level where it can be accepted into a water course. Some of the 'aerobic sludge' which settles out is returned to the aeration tanks, but the remainder is disposed of by adding it to the primary sludge and feeding the combined sludges to the anaerobic digesters.

Most municipal sewage systems use common sewers for domestic and industrial waste-waters and road drainage water. This results in a sewage which contains solid matter suspended in a dilute, aqueous solution. It also results in very large volumes of sewage flowing to the treatment plant. For instance at Mogden (London, England) the sewage flow is over 318 litres (70 gallons) per day per inhabitant of the collection area and the average daily flow is 97 million gallons ($5 \cdot 1 \, m^3/s$).

The sludges fed to the municipal anaerobic digesters are usually comparatively low in solids: about 3–4 % TS is a common figure. Such sludges are thinner than those contemplated for, or used, as feedstocks for agricultural-waste and other digesters planned primarily as gas-production plants, so the gas production per unit volume of digester is lower in the domestic sewage plant than that contemplated for the energy-production plant. In some sewage works sludge is thickened to about 8 % TS before being fed to the digesters, and this should improve efficiency of use of the digesters.

However, even if the feedstock is relatively dilute, gas can be produced in considerable volume from the large sewage-digestion plants. For example, two of the main London (England) sewage works, Beckton and Mogden were producing about 62 320 m^3 and 57 500 m^3 of gas per day in 1972. The ten biggest of the London sewage works produce about 255 000 m^3 gas/day from a population of $7\frac{1}{2}$ million. The total feed to these digesters is about

662 tonnes (dry wt) of sludge per day, or about 0·118 kg per head of population (private communication). Some of the gas is purified and sold, but most of it is used for running dual-fuel engines or gas turbines driving air compressors and generators which provide 80–90 % of the total power required in the sewage works. At the Mogden plant the gas powers continuously 16 engines ranging from 650 HP to 1100 HP which drive 10 compressor sets to deliver about 4·8 million m³ (170 million ft³) of air per day to the activated-sludge plant, and six generators to produce enough electricity for the many pumps and other equipment required to operate the works. Waste heat from the engines is used to keep the digesters at 30 °C through hot-water heat-exchangers.

Even if the digester gas energy is used only to power the sewage system this represents a big saving of mains electricity generated from oil or coal, and some sewage works feed electricity into the national electricity systems. However, domestic sewage might contribute more to energy saving if digestion were more widely practised. Swanwick (1975) stated that some 708 200 m³ (25 million ft³) per day was the digester gas production in Britain, but this was produced by only some 300 out of about 5000 sewage plants in the country. Obviously some of the plants serve only villages and would be too small to consider digestion as part of the treatment system, and in others only comparatively small amounts of gas would be available, but more digesters could be used. Tests are at present in progress on the use in small sewage works of comparatively cheap steel digesters based on those designed for farms.

Cost evaluation of building and running digester plants is difficult to make due to increasing inflation in material and labour charges. On the other hand increasing prices and potential shortages of conventional energy sources tend to increase the value of the digester gas in greater proportion than the increase in digester costs. Natural gas has until recently been comparatively low-priced in Britain and so digester gas was not competitive. The Government's future prices for natural gas, increasing at much more than the general inflation rate, will change the relative costs.

Thompson in 1977 stated that capital costs for digesters ranged from £0·90 to £2·50 per head for thickened primary sludge, and £1·60 to £5·00 per head for unthickened mixed sludge, depending on the population (500 000–20 000). Operating costs for the same range of population varied from 3·0p to 6·5p and 5·5p to 12·5p per head per year. The value of the gas produced was calculated as £4 per tonne of dry sludge solids and equivalent to 14·5p per head of population per year. However, the author believed that a figure of 50p per head per year would be more accurate.

He stated that in the London sewage works the gas has an average methane content of 67% and that the calorific value of the gas is about 22·35 MJ/m³ (6·2 kWh/m³; 600 Btu/ft³). Therefore, 10 m³ of sludge gas is equivalent to 8·25 litres of liquid butane, 6·9 litres of petrol or 6·42 litres of diesel oil. Taking the price of diesel fuel at that time (1975) the value of the gas produced was £2 760 000. The running costs in 1975 of the digester plants were £817 000. The electrical energy from gas being produced at the London sewage plants is equivalent to the electricity used by a town of 150 000 inhabitants. On purely an energy basis there is a case for installing digester and generating plants for towns with less than 50 000 population.

Thermophilic Digestion

The thermophilic range of temperature is from 50° to 65°C and thermophilic bacterial populations for digesters can be developed by incubating various potential sources of bacteria (e.g. sewage, rumen contents, mud, etc.) at increasing temperatures up to 65°C.

Although laboratory and pilot-plant work has been done on thermophilic digestion, full-scale thermophilic digestion plants in operation are very few compared with the mesophilic digester plants now operating. However, the results of laboratory work (e.g. Cooney & Wise, 1975; Pfeffer, 1974) seem to show that thermophilic digestion could have some advantages, although there are also disadvantages.

One thermophilic sewage digestion plant can be found at Playa del Rey, California, USA. Garber *et al.* (1975) reported on this.

Digestion was carried out in a 33·9 m (111 ft) diameter by 9·3 m (30·5 ft) deep tank containing on average a liquid volume of 9650 m³ (2 538 500 gallons). Detention times of between 12 and 24 days and loading rates of 2–3·8 kg VS/day/m³ were used. The digester was heated by injection of saturated steam 1·5 m³ below the liquid surface, and although the steam killed bacteria in the immediate vicinity of the entry point, the water served as an excellent medium for dispersing heat throughout the tank. Gas mixing in this digester has previously been mentioned. Operating at temperatures between 46° and 51 °C had little or no effect on digestion, but when the temperature was raised to 53 °C, within a six week period volatile acids in the sludge tripled in concentration and gas production fell by 25 %.

Thermophilic digested sludge had better dewatering characteristics than that from a mesophilic digester and the thermophilic sludge solids were less gelatinous and more grainy in appearance than the mesophilic solids. The sludge cake produced on a filter leaf or rotary vacuum filter was thicker and drier than that from the mesophilic sludge. However, separation using a basket bowl centrifuge, and simple settling using coagulating agents, were not so good.

Garber and his coworkers arrived at some general conclusions about thermophilic digestion as compared with mesophilic (Garber, 1977). These were that for the same detention time:

1. Volatile Solids reduction was greater in the thermophilic process.
2. Gas production per pound of volatile solids was lower than that at lower temperatures.
3. The percentage of methane in thermophilic gas was greater.
4. The volatile acids content of the sludge was sometimes six times higher than in a mesophilic sludge.
5. Thermophilic plants were more sensitive to rapid temperature change.
6. Thermophilic digested sludge gave more odour problems during the solids separation step.
7. Thermophilic sludge had about 65% of particles passing a 200 mesh screen, whereas mesophilic had more than 80%.
8. It was much easier to mechanically dewater thermophilic sludge.
9. More of the nitrogen was solubilised and then lost in filtration of thermophilic sludge.
10. Thermophilic sludge was adjudged to contain considerably lower numbers of potentially pathogenic bacteria.

Another thermophilic digestion plant (1 million m^3) in Moscow, USSR was described by Papova & Bolotina (1964). Here the operating temperature was 57 °C and conversion from the mesophilic range to the thermophilic range allowed a decrease in detention time from 18 to 9 days and an increase in organic loading from 1·65 kg to 3·5 kg VS/m^3/day. However, total gas production fell by about 3–4%. Investigation of the digested sludges showed that mesophilic sludge retained up to 20% of viable helminth eggs from the input, whereas no viable eggs were found in the thermophilic sludge. As the sludge was spread on agricultural land it was considered of the utmost importance that it was free from viable eggs and pathogens.

It is evident that the thermophilic digestion process can be successfully used, its main advantages being lower detention times, improved dewatering properties of the digested sludge, and better destruction of pathogenic organisms. However, other properties mentioned above, in particular the instability and acute sensitivity to slight temperature fluctuations and the high energy cost of heating digesters and feedstock may prevent it becoming a more widely used process for pollution control and energy production.

DOMESTIC GARBAGE WITH SEWAGE SLUDGE

Domestic garbage (refuse) contains large amounts of inorganic material (e.g. glass, tins, etc.), and plastics which are not amenable to rapid microbial attack, if they can be degraded at all. Garbage as such is not amenable to anaerobic digestion, but after sorting out the inorganic and plastic materials the organic residue consists largely of papers of various sorts and household food scraps: this waste is largely cellulosic, with some fats and proteins and is theoretically digestible.

At present a plant is being constructed in Florida, USA (Pfeffer, 1978, 1980). Here the processing line was designed to handle 100 tonnes of refuse on an 'as received' basis. The refuse is passed through several processing steps before addition to the digester. A trommel screen is used to remove most of the fine inorganic material such as broken glass and sand. A second-stage shredder reduces the particle size to 3·8 cm, and from this shredder the refuse is carried to an air-separation system where the 'light' and 'heavy' fractions are separated—the heavy fraction being rejected. The light fraction is recovered by means of a cyclone system and then taken to a premix tank to be prepared as feedstock for the digesters. Because of the large paper content of garbage the nitrogen concentration may be too low for microbial growth and extra nitrogen may have to be added. Co-digestion of garbage with sewage sludge is one method of increasing the nitrogen content as well as providing an inoculum of digester bacteria.

The two digesters incorporated in the system are each 15·3 m in diameter with a capacity of 1264 m^3. Because mechanical mixing is to be used the digesters have fixed covers. Very little is known about the mixing properties of these garbage slurries, but it was expected that a speed of less than 25 rpm for a 5·2 m diameter impeller would be sufficient. As this plant is still in the developmental stage very few results are to hand, but the gas production rate is expected to be about 8400 to 9800 m^3 per day from the 100 tonnes of garbage and at a digester temperature of 38 °C.

However, about 25 years ago Ross (1954) reported in detail on digestion of garbage and sewage sludge in the municipal sewage works at Richmond, Indiana, USA. Digester capacity of 0·28 m^3 (10 ft^3) per person was provided for a population of 60 000. At the time of writing Ross said that the combined sewage sludge and garbage digestion had been operating for two years. Gas production was about $0.5\,m^3/kg$ VS added and total gas production was about $79\,320\,m^3/month$ compared with $24\,080\,m^3$ when the plant was running on sewage only. The digester gas was used in gas engines to drive blowers for an aeration treatment plant, and a pump unit. It was also used to heat the sewage-works buildings and cooling water from the engines was used to heat the digesters. All the wet sludge was sold as fertiliser. Calculations were shown and indicated that a profit was made from treatment of the garbage.

This, of course, was a relatively small plant and greater difficulties could be encountered in dealing with refuse from a large city. However, the results do show that garbage and sewage sludge can be digested on a commercial scale and not just in the laboratory. Digestion could be a method which, after dewatering of the digested sludge, gives a biologically-stable solid for landfill and reduces the total weight of the original garbage which, untreated, could be open to attack by rodents, by flies and other insects, and by bacteria. The former may cause nuisance and health hazards; the latter odour and gas problems.

OTHER WASTES

While gas production from domestic sewage has been a large-scale process for many years, other feedstocks are now being considered for biogas production. Some wastes which are suitable for digestion and which have been tested, or are being tested, on a commercial scale are those from vegetable-, fruit- and meat-processing factories, farm wastes, and wastes from the fermentation industries.

Factory Wastes
Sewage treatment, either on an aerobic or anaerobic basis, is a microbial process which can be upset by anything which affects bacterial growth. Therefore, local authorities lay down conditions about the discharge of factory wastes into the sewers. In general, two effects can be noted. The factory waste, and this includes farm waste, may be of such strength that overloading of the sewage-plant occurs, leading to a plant effluent well above River Board standards or even to plant failure. In such cases a primary purification of the waste on the factory site must be undertaken to bring the total pollutional load to an acceptable figure, or the discharge of the waste must be controlled so that adequate dilution takes place in the domestic sewage. However, of perhaps greater importance is the discharge

of substances which are bactericidal or bacteriostatic. The effects of detergents, heavy metals and other substances on digestion have previously been considered. But one might note that these substances are also toxic to aerobic-treatment plants, although the fact that some tend to be concentrated in the primary sludges may, in practice, make them have more effect on the digestion plant. In the case of a waste containing toxic substances, these must obviously be removed in some way before biological treatment is possible.

There are, however, many factory wastes in which problems of toxicity do not arise. While these wastes can be, and sometimes are, discharged in controlled amounts to domestic sewers and a charge is made by the authorities for treating them, treatment on site is now being undertaken in many cases, and the possibility of anaerobic digestion and production of energy of use in the factory is a major consideration.

Meat Wastes

The slaughtering of animals and the processing of the meat into tinned products, bacon, pies, etc., produces waste-waters high in putrescible organic matter. The waste contains faeces, urine and possibly bedding, from the stock-holding pens, gut and stomach contents, fatty materials and blood. Some figures given by Hemens & Shurben (1959) give 0·45 kg BOD load per pig unit. (One pig unit is the amount of waste produced by slaughtering one pig, one lamb or one calf, and the amount from one cow or steer is 2·5 pig units.) On this basis, slaughtering of 500 pigs per day would produce a load on the sewage system equivalent to 6650 people.

Digestion is a successful large-scale treatment method for slaughter-house and meat-processing wastes. It decreases and stabilises the solids and removes about 95 % of the BOD. If the waste-water contains large amounts of solids from the holding yards and slaughtering then the possibilities of two digesters could be considered. The sludge from the slaughter-house could be kept separate or settled from the waste-waters and digested in an ordinary stirred-tank digester at a detention time similar to that suggested for cattle or for pig excreta. The larger volumes of low-solids waters from the meat processing could be treated in a contact digester or an upflow sludge-blanket digester.

Steffen (1958) described some results from a meat-processing water, contact digester of which some constructional details have already been described. The gas produced was approximately 80 % methane, probably due to the high fat content of the waste. About 1·5 m³ gas/kg VS destroyed was produced compared with 1·0 m³ gas/kg VS destroyed for domestic sewage. The digestion process was extremely stable and could easily be shut down and restarted.

In 1975, in the Philippines, a digester was built to control pollution from piggeries and meat processing and canning plants (Obias, 1976). The gas was used to generate electricity, run engines for deep-well pumps, run gas refrigerators, heat drying rooms, and for smoke generators and gas cookers. A detention time of 30 days was used in a single-stage tank and the effluent was stored in lagoons for 30 days before being used as fertiliser. The use of the digested sludge as fertiliser enabled three crops a year to be grown.

Fruit, Vegetable and Other Crop-processing Wastes
This type of waste is high in organic matter and should readily digest, as laboratory experiments have shown. However, large-scale plants are not, as yet, numerous.

In Malaya the economy is largely dependent on the production of rubber and palm oil and it is the palm-oil effluent that has given rise to major pollutional problems (Wood *et al.*, 1979; Morris, 1980). From each tonne of fresh fruit bunches processed, approximately 0·2 tonnes of palm oil is obtained, together with 0·65 tonnes of highly-polluting effluent with BOD values ranging from 20 000 to 30 000 mg/litre and COD values between 60 000 and 80 000 mg/litre. Thus the palm-oil industry was producing pollution equivalent to 1·25 times that of the total population. Legislation, backed by heavy fines, has forced companies to deal with the problem and after pilot-plant investigations, large-scale, single-stage, steel digesters have been built ranging from 1500 m^3 to 3500 m^3 capacity. They have at present been operating in the mesophilic range with detention time of 20 days. BOD and COD reductions were 87 % and 72 % respectively, and although large quantities of gas were being produced no use was being made of it at the time of writing (Morris, 1980).

In New South Wales, Australia, experiments have been carried out using wastes from a fruit canning factory (*Rural research*, 1979). Apricots, pears, apples, peaches and citrus fruits have all been digested in a 25 m^3 digester, which is 1/20th of the size of digester needed to treat the factory's full output. Dried orange peel has also been digested and it has been estimated that a full-scale digester in a processing plant with an annual output of 30 000 tonnes (wet wt) could produce about 2·7 × 10^6 m^3 of gas per year (60 % methane, 40 % carbon dioxide).

Wastes from potato-processing, pea-processing and other effluents are ideal for digestion. However, as with most fruit-processing factories, there is a major difficulty in that the production of wastes is seasonal. Presumably a factory could process other fruits and vegetables at different seasons and the digester flora should easily acclimatise from one waste to the other. However, optimum conditions for digestion of different wastes vary and difficulties could be encountered in varying digestion plants geared to producing energy for factory use.

Two plants using contact digesters have been made to treat wheat-starch-gluten factory wastes. The larger plant in Bordeaux has a loading of 5000 kg BOD/day and produces 4080 m^3 of gas/day. The calorific value of the gas is given as 6040 kcal/m^3 (7·02 kWh/m^3), so the daily gas energy production is 2·866 × 10^4 kWh. The energy input to the digester plant is 38·5 kW and the digester operating temperature is 33·5 °C. The BOD of the waste-water is 13 980 mg/litre and the COD 22 370 mg/litre and these values are reduced by 94 % and 96 % in the digester effluent (Donnelly, 1978). Other large pilot-plant digesters of this type are successfully treating creamery and other factory waste-waters in Britain.

Donnelly states that the power input to the digester is 90 % less than that to an activated-sludge plant treating the same effluent, so that while the capital costs of the digester and an aerobic plant are similar, the running costs of the digester are lower and the digester actually covers these costs, and makes a profit, with the gas output. Scammell (1976) also said that the cost of building modern anaerobic contact digesters to deal with these waste-waters of low solids by high BOD content is comparable with that of the classic aerobic systems, but the running costs are negligible. The economic advantages of the anaerobic method can be applied to all biodegradable effluents. The low running cost, in terms of energy, for digesters treating other wastes as compared with aerobic plants giving the same reduction in BOD has been pointed out a number of times (e.g. Hobson & Robertson, 1977; Mills, 1977) and is mentioned elsewhere in this book, as is the comparative stability of digester plants.

A large-scale (200 m^3) upflow anaerobic sludge-blanket digester (Fig. 8.1) has been built to treat waste-water from a beet-sugar refinery and is reported to be working satisfactorily (Lettinga *et al.*, 1980). All waste-water from the refinery is treated with about 95 % efficiency in a liquid detention time of six hours. No excess sludge is to be disposed of. Any excess produced is sent back to the semi-anaerobic pretreatment plant. Enough methane is produced to satisfy the needs of the industrial lime-oven of the refinery where lime is produced to precipitate the raffinose from the sugar syrup and allow higher sucrose yields (Nyns, 1978).

The simple anaerobic filter has been successfully used to treat starch-

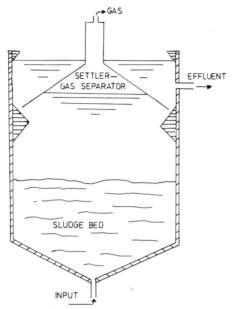

FIG. 8.1. Schematic sludge-bed digester. The effluent take-off would go to a weir or other system for balancing gas and liquid pressures in the digester.

gluten waste from a flour mill (Taylor, 1972). The waste input was treated for adjustment of pH and temperature and passed through two upflow filters. Flow rates were 590 m³/day with a filter loading of 108 kg COD/28 m³/day (3·86 kg COD/m³/day). Hydraulic detention time was about 0·92 days. On average, COD removal was about 64% and gas production was estimated as 850 m³/day. The advantages of the anaerobic filter system were listed as:

1. Low production of biological solids;
2. High treatment efficiency;
3. Low capital cost;
4. Methane production;
5. Low nutrient requirement;
6. Low operating costs;
7. Minimal attention required;
8. Inherent long solids-retention time.

Fermentation-industry Wastes
Fermentation industries, that is factories producing alcohol, produce large

volumes of waste-waters which contain soluble carbohydrates, etc., as well as solid residues containing cellulose, hemicellulose, proteins, and other substances. These wastes are ideal for digestion as the results of pilot-plant studies show. So far, full-scale plants have been designed to use digestion as a means of reducing pollution with the gas used to reduce the costs of the process.

Information on large-scale plants is sketchy, but Ono reported in 1965 that there were ten factories in Japan producing alcohol from molasses, sweet potato or grains. Each of these ten distilleries used digestion for waste treatment. Of the seven plants using mainly molasses, four had mesophilic digesters and three had thermophilic plants. The thermophilic digesters operated at higher loading rate than the mesophilic. Gas production was in the range $0.49\,m^3$ to $0.62\,m^3$ per kg TS added. The gas was used with coal to fire boilers.

Wastes from a Japanese alcohol distillation plant with a BOD content of between 20 000 and 50 000 mg/litre underwent thermophilic digestion at $53°-54°C$ and detention times between four and seven days. Maximum organic loading used was $16\,kg\ VS/m^3/day$. Up to 89% of the BOD was removed and the gas contained $50-60\%$ methane.

Warm effluent ($50°C$) from an 'alcohol from whey' distillation plant in Denmark is used as a substrate for biogas production. An anaerobic contact digester is employed with a vacuum degasifier prior to the separation tank. The gas ($63\%\ CH_4$), which is practically odourless due to the very low sulphur content of the whey, represents an energy value equivalent to $1.8\,kg$ fuel oil/m^3 of whey waste. As there is no need to heat the digester all the gas is available for use at the distillation plant where it replaces $17-20\%$ of the fuel oil required (Reesen & Stube, 1978).

Two or three contact digesters like that previously described for the wheat-gluten plant have recently been set up in Britain to treat distillery wastes.

Waste-waters from other fermentation plants producing antibiotics, single-cell protein, etc., should be amenable to anaerobic digestion.

Rudolphs & Trubnik (1949) built a two-tank digester which operated for five years and was to be a pilot-plant for a full-scale digester processing yeast-production waste-waters. This plant, although described as a pilot-plant, is big enough to be included in this section. The preliminary laboratory work has been described in a previous section.

Each tank was of 16 000 US gallons ($60.5\,m^3$) capacity, the top being a gas collector and the maximum liquid volume being 14 000 US gallons ($53\,m^3$). Three other liquid volumes could be used by operating the appropriate set of inlet and outlet pipes. Heating to 85 °F (24 °C) was by internal steam coils. The digesters could be used as two single-stage digesters in parallel or in series as a two-stage digester. Six months build-up of digestion from a digested sewage sludge inoculum was used before full loadings were applied.

The digesters were operated as a type of combination of upflow sludge blanket and feedback process. The raw waste was fed continuously into a distributor at the bottom and passed upward through the sludge. The sludge collecting in the effluent settling tank was returned periodically to the digesters.

The two-stage digester gave higher BOD removals at all loadings than the single-stage, with some 78 % of the overall BOD removal taking place in the first stage and 22 % in the second stage. However, at very low BOD loadings the percentage BOD reduction was similar in each tank. The authors suggested that the results showed that the system did not behave according to the conception that a two-stage system would treat easily-digestible materials in the first tank and more resistant materials in the second. The two-stage system was behaving much as the theory given in an earlier section, but, obviously, the sludge return would cause some divergence from this theoretical model. Volumetric detention times of 3·7 to 1·4 days were used and did not affect results. However, the sludge solids (bacterial) detention time would be much greater than this. Gas production was about 7·5 ft³/lb VS added on average, but varied with BOD loading rate. Volume per unit of VS added increased as loading rate increased up to about 0·07 lb VS/ft³ digester/day and then decreased, but carbon dioxide content increased from 18·2 % at the lowest loadings to 32·7 % at the highest. The hydrogen sulphide content (the wastes have high sulphate) increased from about 2·0 % to 3·9 % with increased loading. The rest of the gas was methane. The raw-waste had a pH of about 5 but was easily neutralised by the digester contents and so predigestion neutralisation of the input did not affect digestion. Shock loadings, and no loading for at least a week, did not affect digestion.

This process worked well and in combination with a trickling (aerobic) filter for final polishing of the settled digester effluent gave very good purification of the effluent, as well as biogas production.

Farm-animal Wastes

Farm animals produce large amounts of faeces and urine. If the animals are kept outdoors then this waste returns naturally to the land to provide an excellent fertiliser. In countries with cold winters animals cannot be left outdoors all the year round, and during the winter the disposal of the excreta may be a problem. Because of weather conditions, even if land is available for manure-spreading the waste may have to be stored for long periods.

The increasing demand for meat, dairy products and eggs has led in many countries to intensive farming of cattle, pigs and poultry. In this type of farming system the animals are kept indoors for all, or in the case of some dairy units, for much, of the year in sheds, or outside in feedlots, in dense populations. Many intensive farm units of this kind do not have the land for fertiliser spreading of the animal excreta and it has to be put to land in 'sacrifice areas' in amounts far higher than any fertiliser use, or transported to other farms. The storage and eventual disposal of this waste can, and does, raise problems of pollution, especially when the livestock units are near houses, villages or towns or water courses or reservoirs of various kinds. The problems of waste disposal on the intensive unit have been added to by the use of 'slurry' excreta-collecting systems with minimal animal bedding. This can make waste handling easier on the unit, but produces a waste which is more likely to cause odour and run-off pollution than the traditional 'farm-yard' manure of excreta plus straw or other bedding.

The total amounts of animal waste produced are described in Appendix 2, but some comparison with domestic sewage may be made here. The population of Great Britain is about 56 million. The population of cattle has been estimated as 14·5 million, of pigs 9·0 million and of poultry 144·1 million. Since the excreta of one cow is roughly equal to that of eight humans, one pig to four humans, and eight hens to one human, it can be seen that the total waste of animals, of which a growing proportion are intensively housed, greatly exceeds that of the human population. This animal waste represents a source of biodegradable material and anaerobic digestion offers a means of treating these strong wastes to reduce pollution and odour as well as to produce gas. The digested waste, as shown before, retains the fertiliser value of the original while being much better to store and spread, and the advantage of digestion over other forms of treatment is that the digestion interposes a pollution-control step between production and utilisation which is energetically cheap to run and which can on suitably-sized plant cover not only this energy but produce excess energy of use on the farm.

The digester for waste from two or three cows is a practical impossibility so far as producing energy for the cold-country farm is concerned. Once a digester becomes above the 'few cow' size, it then needs automation in running, especially in countries where labour is expensive. This automation adds to the energy deficit of running the digester, but this does not increase in proportion to the size of the digester. So there is a minimum size of digester which can cover the costs of automation. A warm climate will increase the useful energy balance of the large, automated digester as little, or no, energy will be needed for digester heating, but the energy cost of the automation has still to be considered.

These balances of energy input and output mean that the automated digester is being considered as an energy-production (and pollution-control) system to meet, or help to meet, the needs of the intensive farm unit of 2000 pigs, 100 cattle (or their equivalent in poultry) and above. In the case of contributing to energy needs outside the farm the units being considered are in tens of thousands of pigs and cattle, and up to the hundred thousand or more of either.

Small Digesters

Despite our annual 'energy crisis' the industrialised countries per capita consumption of energy exceeds five tonnes of coal equivalent per year, while in North America it exceeds 10 tonnes. It is the developing countries, especially those in subtropical and tropical Asia and Africa, where the average per capita energy used is less than 0·3 tonnes of coal equivalent annually, that are facing in many ways a worse, and worsening, energy crisis. Firewood, the principal source of energy in rural areas is now very scarce and expensive (Openshaw, 1974). The ever-increasing price of imported crude oil is forcing poor countries to spend most of their export earnings on keeping their fuel supplies at a constant level. Most of the fuel in these countries is used for cooking and lighting, and biogas generation using simple digesters could not only provide fuel but improve sanitation and health and crop yields when the digester effluent is spread on land. It has previously been said that it is difficult to obtain information on how many, and how well, most of the small biogas generators are working in the poorer countries, and Pyle (1980) states that only in two or three countries has biogas production made any impact, and nowhere, with the possible exception of China, does biogas make a significant contribution to either energy production or recycling of organic matter.

Why is this the case? Is it lack of technical know-how, poor digester design, lack of adequate technical help and follow up, or are the problems of survival already too much to cope with for these people of the poor nations?

That even the simplest digester needs some technical expertise in construction and a reasonably regular running schedule has already been pointed out and many of the difficulties probably stem from badly-engineered digesters even though the basic principles of digester design are fairly well known. Problems have arisen in digester construction, sealing the digester, mis-aligned gas-holder guides, misplaced feed pipes, poor control over feed input, corrosion and lack of maintenance. Major difficulties can also arise through inadequate mixing and poor temperature control, especially at low ambient temperatures, and where night and day temperatures differ considerably. Another source of poor performance could be in poor feeding of the animals. Poor feed will lead to little and poorly digestible faeces.

These problems should be overcome eventually through better instruction of users and builders and the trend towards 'village' rather than 'peasant-farm' digesters previously mentioned will help.

As described before, the digester designs in the poorer, hot countries are many, but in the following section a brief summary of known results is given to show what is being done in different countries.

India: Much of India has a high ambient temperature of about 36 °C (97 °F) which should be ideal for digestion. The majority of India's digesters are of the type already described. Plants ranging from 2 to 140 m³ capacity have been built. The smaller domestic units are unheated and mainly unstirred and operate with a detention time of 55 days. Gas production, at about one-quarter digester volume a day, is disappointing and this inefficiency is probably due to some of the reasons mentioned earlier. The Indian Government have a programme to install some 500 000 digester plants over the next few years.

Korea: About 29 000 small plants are reported to have been installed. Designs are very similar to the Khadi Village Industry Commission (KVIC) digesters of India. Because of the low winter temperatures the plants are effective for only about six to seven months in the year.

Nepal: From about 1975 some 200–300 plants have been built with about 80 % success rate. Finlay (1978) points out the now familiar problems of the plants (mainly based on some KVIC designs, although some digesters have been built to the Chinese designs). Finlay states that their financial viability is only marginal and has also pointed out the great managerial problems involved in carrying out a major digestion programme in a poor country.

Taiwan: The design of Taiwanese digesters has been discussed in a previous chapter and the Taiwanese appear to have met with some success in their digester systems. Chung Po (1973) said that 'if operated properly' a unit for a 20 pig farm should provide fuel for cooking for a family of twelve. The use of the digested sludge for growing algae for animal feeding is part of the programme.

China: Chinese digester design was covered earlier and the digester programme seems to have become successful in a relatively few years, although it is difficult to get exact information. Because of the simple design, digester maintenance is said to be minimal (Khandelwal, 1978). More than 100 000 technicians were said to have been trained in digester construction by 1975 in a mass training programme and it is probably to this type of organisation in training of people in the construction and use of digesters that success is due. The numbers of digesters built and running is difficult to determine. A figure of five or seven million has been mentioned, but Smil (1977) said that it had been estimated that 43 million had been built.

In the book *A Chinese biogas manual* an interview given in late 1978 by Mr Liang Daming, the Head of the Biogas Works Office, Shachiaho Commune, Guangdong Province was recorded and this shows the kind of work going on. In 1976 three people, including Liang, from the Commune were sent to Sichuan Province, where the digester programme originated, to be trained in digester construction and operation. The three returned home and built five demonstration digesters of which one failed. Because of the high water table they had to develop a new method of construction of the digester tanks. A further 70 technicians were trained to spread the techniques. Some months after the start of the programme 100 digester tanks had been built and two years after in late 1978 1640 digesters had been built, 1300 by individual families. All of these, except for the one original digester, are apparently working. They range in size from 5 to 7 m³ for individual families and 30 to 40 m³ for collective units. The former generate gas for cooking and lighting and the latter generate electricity, but fertiliser from the digested sludge is an important product.

The original 70 technicians now work full-time in the province checking on safety and management, maintenance and repairs and generally helping any digester owner who has problems. Gas production is 0·15 to 0·20 m³ per m³ of digester volume at an ambient temperature of about 25 °C, and it seems that above 30 °–33 °C about 0·3 m³ of gas can be obtained (this latter temperature is indefinite as there is an inconsistency in the temperature figures given). The digester should be loaded at least every three days and preferably every day. Roughly 50 kg of slurry (about 20 % organic matter, 80 % water) is required per day for each cubic metre of gas produced. Equivalent sludge is removed after feeding. The digesters are emptied once or twice each year to remove accumulated fibre and to make repairs. In winter the ambient temperature goes down to 5 °C and too little gas is produced, so the digester is covered with rice straw as insulation and input is added more frequently (presumably to make up for the lower specific gas production).

The cost of Chinese digesters is impossible to estimate. The construction materials are cheap, but vary according to local situation, but a lot of labour goes into the construction, running and maintenance, and the value of this labour in a peasant farm or commune economy cannot be translated into 'western' terms.

The Chinese have no special knowledge of digester running and construction; attention to detail and organisation and hard work seem to be the reasons for their success. This is a pointer for other digester programmes.

Large-scale Digesters

The design and construction of large-scale, automated farm-waste digesters has already been considered. It was pointed out that although most are on the single-stage, stirred-tank principle, detailed design varies considerably because this generation of digester is developmental. For this reason long-term results of digester running are not generally available. The digesters have not been in operation for long and they are still subject to modification as difficulties appear. What may be said is that no microbiological difficulties have been found. The digestion can be started relatively easily and once started the microbial system is stable. Gas production is as determined from small-scale experiments. The problems that arise are mechanical, in obtaining regular inputs, preventing scum formation and blockages in pipes and overflows, tuning engines correctly, and so on. These difficulties, which can be, and are being, overcome, have to be set against solutions which use little energy and cost little.

The many 'engineering' and 'economic' models which have been made using data from small-scale experiments all show that pollution control and energy production by digestion of farm wastes is worthwhile on various scales. But the model system and even the detailed design drawings of a particular system cannot, at the moment, foresee the kind of problems that develop in practice. This is why we now need practical experience in building and managing digesters, not just further modelling.

Even when digester designs have been more perfected there will, for the on-farm digester, still be a need for study of the particular site and its waste output and particular difficulties. No two digester systems will be exactly alike. Farm wastes and farms are much more varied than municipal sewage and sewage systems, and the package-deal system for a farm of *x* number of animals such as is now offered in sewage treatment for *y* head of population will not be possible. That is not to say that basic designs and parts of digesters cannot be common to many sites, but details will vary and, for instance, the same detailed design of digester could not be used for cattle and for pig wastes; the properties of the wastes are so different.

It should also be remembered that even the most perfect digester will in practice have occasional difficulties with blockages, etc. The modern sewage works, in either aerobic or anaerobic plant, does not run perfectly all the time. What one has to aim for is to cut the stoppages and maintenance requirements to a minimum.

To give some idea of the present situation and results the following section discusses briefly the position in various countries.

Great Britain: One of the 13 m³ digesters on the Aberdeen site has, as was previously mentioned, been operating on pig slurry for some seven years. Design features were described in a previous chapter and it was pointed out that this particular design could not be greatly scaled up. Nevertheless detailed monitoring has shown that its performance has matched that predicted from pilot-plant experiments and in respect of gas production at higher solids concentrations it has been slightly better on average than the small-scale plant. This was discussed by Hobson *et al.* (1979) who gave an analysis of results from a number of years of running. Mills (1977) also gave some results when comparing the performance of the digester and an aerobic treatment plant running on the same piggery waste. As the plant is experimental, pig slurries of different Total Solids have been tested. A few general results of the earliest running were given by Bousfield *et al.* (1972). Some earlier results at longer detention times during start-up of the digester were given by Robertson *et al.* (1975) but a ten-day detention time was decided from pilot-plant results to be optimum and this time has been used for many years with an operating temperature of 35 °C. Gas production has averaged 0·30 m³/kg TS (containing 70 % VS) added to the digester over a range of TS from 2 % to *c.* 6·5 % (the latter the highest TS that can be handled with the small pumps), and at 6 % TS input, about 24 m³ of gas (i.e. nearly 1·9 digester volumes) of 69 % methane content are produced per day.

Reductions in BOD and COD between input and output (whole sludges) have averaged about 80 % and 50 % for the various loadings of TS. Reduction in TS has averaged 40–45 % although because of a rather higher percentage of organic material in the higher TS slurries, TS reduction in these has been *c.* 50 %. VFA reduction has been about 75 % on average, but this percentage reduction varies somewhat with input as the residual VFA level remains about the same if the detention time is constant. Ammonia nitrogen is little changed, sometimes being slightly higher, sometimes lower than the input. The digested sludge has little or no odour, and has been eventually land-spread as fertiliser.

At times, mixtures of piggery waste with cattle and poultry wastes, and piggery waste with waste grass-juice from a grass-pressing experiment have also been successfully digested.

Although some 'engineering' problems have arisen at times due to blockages (and initially some experimental changes in mixer design were made), biological problems have been nil. Since its initial start-up from a municipal digester-sludge seed, the digester has never had to be reinoculated in spite of stoppages in feeding and heating; removal and replacement of the digester contents; varied loading rates and the above-mentioned changes in feedstock. A low loading rate for a few days, after, when necessary, a short period of warming up to 35 °C, has been followed by full activity of the digester flora. Periods with no loading and no heating, with the digester contents in another tank have extended to a number of weeks on two or three occasions, when major alterations were being made.

The digester has recently been turned over to digestion of dairy-cow slurry, using the piggery-waste digestion as an initial seed (inoculum), but at the time of writing the digestion is still being built up to full loading rate. The pig slurry inoculum has, however, been entirely displaced by the cow slurry and gas production is continuing at a rate commensurate with that calculated from the pilot-plant experiments.

A larger, stirred-tank digester was made some two to three years ago by a commercial firm (Davies, 1979; some operating details reported by Hobson, 1979a). This has been taking waste from an experimental unit with a variety of animals and was running on a 6 % TS slurry of pig, chicken and human waste and food-laboratory waste. During the first year, although not running at full input, the gas was heating the digester and supplying 220 000 Btu/h (6·45 kW) over an 11 hour day to a system also utilising tail-gas heat from an incinerator to heat buildings and provide hot water on the site.

Some commercial, stirred-tank digesters of about 11 m^3 capacity and made of glass-reinforced plastic have been operating successfully for two or more years on small farms. Where necessary two or more of these digesters have been run in parallel to take larger amounts of pig or cattle waste. Gas is used for digester heating and space heating (Chesshire, 1978).

A number of large-scale farm digesters (c. 400 m^3) have been recently built, or are being built, in Britain, but none have been running long enough for comprehensive results to be obtained.

A 390 m³ digester taking slurry from a 500 sow unit has been installed in Yorkshire (Chesshire, 1978) and is producing gas, and eventually this will run as gas-engine generator. Pollution control was a major factor in installation of this digester and the farmer has installed an irrigation system for use of the digested sludge as fertiliser to take the place of the previous two tractors with slurry tankers. He has reduced his spreading costs from £2·50 per 4·54 m³ (1000 gallons) to 55p per 4·54 m³ as well as reducing pollution problems, and local farmers are now keen to take the almost odourless effluent (*Farmers Weekly*, 1979).

The below-ground digester which was described earlier is taking slurry from 320 dairy cattle. It has a nominal capacity of 227 m³ (50 000 gallons) and is designed to run at a 14-day detention time. Some initial problems in crust formation were overcome by alterations to the gas stirring and the digester has been producing gas to run a gas-engine generator of 35 kVA for 18 hours a day (C. E. Dodson, 1978; private communication).

Ireland: An anaerobic filter taking the supernatant liquid from settled waste from 1000 pigs is being tested. Design details have been previously mentioned. At a liquid loading of about 4·5 m³/day, an operating temperature of 28°–30 °C and liquid retention times between 12 and 36 hours, gas production was said to be 89 m³ per day, while COD removal was reckoned to be 90 % (Newell *et al.*, 1979). The methane content of the gas was also said to be 80–85 %, a much higher percentage than that normally obtained from stirred-tank digestion of piggery waste.

A 10 m³ stirred-tank digester has been running for a year or two in connection with the previously mentioned plant producing a urea-formaldehyde encapsulated digested sludge. This digester is taking piggery waste (B. Gavin & J. Watson, private communication).

France: Development here has tended to concentrate on the batch-type digesters for solid wastes previously described, and gas productions are said to be about 0·4 m³/kg TS added, over the production period of about 30 days for one set of digesters using cattle waste with bedding. Another set of solid-waste digesters with a feedstock of piggery waste plus large amounts of bedding are said to have gas production rates of less than 1 m³/m³ digester volume per day (quoted by Wheatley, 1978).

However, stirred-tank digesters taking animal-waste slurries will be commercially built in France. One digester of 1125 m³ capacity has been built and is now producing gas, but is not at the time of writing up to full production rate (D. Evers, personal communication). The gas will be used to generate electricity for the full waste-treatment and fertiliser production plant previously described.

Holland: Most experience is being gained with the sludge-bed digesters for factory wastes previously described, but a 300 m³ stirred-tank digester has been built to handle waste from 4500 pigs, and the economic assessment is that it is a viable process.

Switzerland: A number of digesters have been built and have started operation. For example, it is said that a 200 m³ digester is operating continuously at 35 °C and a retention time of 10–15 days and producing 160–200 m³ of gas per day from cow slurry.

Sweden: The only information available on large-scale farm digesters comes from one farm which has three 100 m³ digesters, each producing 65 m³ gas per day from cow slurry.

Finland: A digester of 22 m³ volume of novel design with a horizontal, cylindrical tank with input at one end and output at the other and stirred by a three-blade rotor on a shaft running centrally along the tank, has been working for some months at a small pig farm. The gas is used in a boiler for hot-water heating of the piggery. Because of the severe winter climate the whole digester is in a building. This digester was produced commercially and there are plans to build further digesters in 55 and 85 m³ sizes, although this latter, for mechanical reasons, is probably the largest that could be built to this design.

Denmark: Of all the Scandinavian countries Denmark seems most strongly committed to anaerobic digestion. The Danish Government has set up a two year programme costing about £300 000 to investigate digestion and the plans include three full-scale plants (Grøn, 1980). Each plant is of a different design. One digester is a completely-mixed, single-stage digester of 270 m³ capacity and will operate on slurry from 1700 pigs. The second is a completely-mixed, batch or continuous two-stage digester, each vessel having a capacity of 180 m³. It will digest slurry from 96 cows, 500 pigs and 100 sows. The third digester is a plug-flow reactor system with two digesters in parallel. Each digester has a capacity of 200 m³ and is built from concrete blocks with a flexi-PVC gas-holder top. The feedstock will be waste from 150 cows plus progeny. The whole programme is expected to be complete by mid-1980 when all the digesters should be fully operational.

USA: A number of farm-scale digester plants have been built in the USA and are, as elsewhere, in the developmental stage. Because America has much larger cattle-raising operations than have European countries, attention there has turned towards the very large digester plant capable of producing pipeline gas for a whole town or area.

One of the largest plants can be found in Guymon, Oklahoma, where two digesters each with a capacity of $7575\,m^3$ have been installed (Meckert, 1978). The plant is designed to process about 500 tonnes of manure per day from a 100 000 head cattle feedlot. The manure from the feedlot is slurried and fibre extracted by screening for use as cattle feed. The screened slurry is fed to the digesters and the gas produced ($45\,325\,m^3$/day; 60 % CH_4) is scrubbed to give pipeline quality methane. The digested effluent is centrifuged to produce a high-protein sludge and a liquid fertiliser. At the time of the report quoted the plant was still in the testing stage and while no difficulties had been found on the biological side, some mechanical problems had occurred but were being overcome. The construction of this plant has been mentioned previously.

Large-scale fermenters digesting poultry waste are very few. However, one such digester was described by Converse *et al.* (1977). The plant was a stirred-tank, $96.5\,m^3$ digester running at 35 °C, loaded from 1·6 to 2·0 VS kg/m^3/day and with detention times of from 30 to 52 days. Gas production was from 55 to $74\,m^3$/day. The methane content of the gas varied from 55 to 63 %. Net energy output ranged from 18 to 75 % of gross energy. Ammonia concentrations varied from 6200 to 8000 mg/litre and these high values may have been the cause of the low gas productions; less than one digester volume per day. Results from the authors' pilot-plant studies on poultry excreta give gas production of three digester volumes per day at digester ammonia concentrations of 2500–2800 mg/litre (Bousfield *et al.*, 1979).

An interesting large-scale comparison of plug-flow and completely-mixed anaerobic digesters fermenting cattle waste was described by Hayes *et al.* (1980). The construction of the plug-flow digester has been previously described. The capacities of the two digesters were similar, the plug-flow $38\,m^3$ and the stirred-tank $35\,m^3$, and they were compared under the same conditions of temperature, detention times and loading rates. In all cases the plug-flow reactor gave slightly better results; e.g. at a 15-day detention time gas production (vol/vol of digester) from the plug-flow digester was 2·33 and from the stirred-tank 2·13. Gas composition, at 55 % methane, was the same for both. The operational data showed that the plug-flow digester maintained higher rates of solids conversion to gas.

Economic assessment suggested that biogas could be produced on dairy farms of from 25 to 500 cows at a stable cost comparable to current US prices for propane and fuel oil. Although a rubber-bag plug-flow digester might be thought to be cheaper to construct than a stirred-tank, comparisons in Britain suggest that the rubber-bag digester would be equal to, or even greater than, the cost of a steel-tank digester. However one wonders how far this particular design could in practice be scaled up before mechanical problems ensued.

Other countries: In many other countries programmes for development of commercial-scale digester plants using farm wastes are under way or beginning. For instance, Dominica is experimenting with cattle-waste digesters and there the hot climate helps greatly in the overall energy balance. Excreta from feedlot cattle on feeds containing molasses and bagasse do not seem to present some of the handling problems posed by excreta from animals on long-fibre European diets. Brazil has embarked on a digester programme as well as its fuel-alcohol programme.

General: As can be seen there are various possible designs for farm-waste digesters. In addition there are variations on the ancillary equipment, e.g. fibre separators for feedstock and digested sludge, centrifuges and driers for fractionating and processing sludge solids, and purifiers for gas. The type and construction of digester depends to some extent on size of unit and some ancillary equipment is economically and mechanically suited only for the very largest units.

Domestic sewage digestion has been practised for years, but even in this field improvements are possible, and efficiency in energy production is now more than ever being aimed at. Of the 'newer' substrates for digestion, farm-animal wastes hold most promise. The waste is concentrated in a small area, it is digestible, and because of its high solids content gas production per unit volume is high. Small-scale tests and calculations and modelling based on these suggest that farm-waste digestion can be energetically and economically sound. The present generation of digesters have gone a long way towards proving this. But there are still practical problems to be overcome, and some types of waste are easier feedstocks for digestion than others. What is now needed is running experience over a few years with commercial-scale plant of different kinds with different feedstocks, so that the theoretical predictions can be shown to be fulfilled.

'Energy-crops' and Crop Residues

Small-scale experiments have shown that vegetable matter can be digested to produce methane. There are three ways of looking at the use of vegetable matter in digesters. The vegetable matter could be used in the form of a seasonal waste to augment gas production from a digester normally running on some other feedstock. This was the view taken for the authors' experiments with potatoes described in another section. Damaged and undersized potatoes, unfit for human consumption, are always a waste product from potato harvesting and are sold for cattle feed or other purposes. The experiments were to test whether it would be worthwhile using this seasonal waste as an additive to farm digesters to boost gas production in winter when gas is most needed. Large amounts of vegetation are generally left after harvest of some human or animal food crop. This vegetation could be used as a feedstock for digesters. To overcome the seasonal production of the waste the vegetation could either be stored or different kinds of waste could be used at different seasons. Finally, crops could be grown purely as digester feedstock for large-scale gas production.

Obviously, if the crop can provide a foodstuff as well as digester feedstock then this should make the project more economically sound. But a difficulty about this is that the food portion of the crop is the part most easily anaerobically digested and likely to give the largest yield of biogas. The residues are very often lignified and fibrous and digestible only with difficulty (*c*. straw digestion). However, the most important point in discussing the use of crops and crop residues is the yield of crop per unit of ground and the number of crops that can be obtained per year.

Most of the work on crops is at the moment in the modelling stage, based on small-scale experiments. Such models (e.g. one for grass, Hobson, 1979*b*) can show the difficulties in trying to obtain biogas equivalent to any large fraction of the total energy usage of a densely populated country, with a cool climate and relatively low crop yields, such as Britain; but this model did show the enormous overall energy benefits obtained by recycling the digested crop residue as fertiliser in place of the chemical fertilisers now extensively used. On the other hand the models can show that in some areas with high crop yields and the possibilities of more than one crop a year (and ambient temperatures which decrease energy used for digester heating), gas production from crops could make a useful contribution to energy needs of the country as a whole. Brazil, for example, has huge areas of well-watered underdeveloped land as well as a warm climate, and a large ratio of land to population.

New Zealand has a good climate for growing crops, it has a small population for its land area and the costs of crop production are comparatively low. Eighty percent of New Zealand exports are agricultural produce and this earns currency for the purchase of petroleum amongst other things. Because of low world demand and import tariffs imposed by many countries New Zealand has been unable to sell more agricultural produce to make up for the increases in fuel oil costs, but it has the capacity to increase its crop production by about 50 %, so these crops could be used for energy production and help to solve some of New Zealand's problems.

Stewart (1980) states that biogas production from crops is economically competitive with all premium fuel available in the South Island of New Zealand, but that it is not yet competitive with the natural gas and propane available from the gas fields of the North Island. Using a $45 \cdot 5 \, m^3$ single-stage, stirred-tank digester, loaded with 10 % TS vegetable slurry and with a detention time of 15 days, he showed that gas production was about $136 \, m^3$/day. The gas had a 70 % methane content and only about $19 \, m^3$ of the gas was used per day for digester heating, leaving a daily surplus of $117 \, m^3$. Gas yield is in the region of $0 \cdot 45 \, m^3$/kg TS added, which is about double that obtained from cattle waste. On a commercial scale the use of relatively small digesters was still envisaged, the gas being used to replace petroleum fuels by the farmer, a group of farmers or the local community. The detailed calculations based on the prototype digester running included a profit margin for the crop producer. In the longer term (20 years or more onwards) larger scale production might be used.

The crop (for example maize) is grown near to the digester and stored, after harvesting with a fine-chop harvester, as silage. This silage is then drawn on as necessary and mixed with some of the effluent from the digester in a premix tank and loaded into the digester. The digested sludge supplies the nitrogenous fertiliser needed by the crop, so only phosphate chemical-fertiliser has to be bought in. Crop yields are two or three or more times those in Britain.

The gas can be used as it is, or scrubbed and compressed for vehicle use.

Another source of crops for large-scale biogas production is the sea. Small-scale tests have shown that seaweeds will digest and models have been made of seaweed 'energy-farms', in suitable climatic areas. Some large-scale work on seaweed harvesting and other aspects is being done. For instance, the feasibility of kelp farms off the coast of California is being looked into (Kohn, 1978). One kilogram of dried kelp yields about $0 \cdot 37 \, m^3$ biogas and with a production of 114 tonnes of weed/acre/year, methane production is expected to be about $3500 \, m^3$/acre/year. Nutrients for the kelp will be provided by the upwelling of nutrient-rich water from below the farm, and when conditions are right the Californian Giant Kelp grows to very large size, very quickly, with an excellent regenerative capacity after being harvested. Economical harvesting of the weed, in terms of money and energy, would appear to be the biggest factor in practical application of seaweed farms.

Landfills

It was said earlier that methanogenesis occurs in wet soils and muds where diffusion of air is limited and the anaerobic bacteria can develop into active populations. A large pile of domestic garbage in a climate with a reasonable temperature and rainfall offers a good site for such bacterial action. The garbage compacts, thus slowing diffusion of air into the mass and metabolism of aerobic micro-organisms quickly uses up the oxygen and can also tend to warm the garbage. The damp mixture of papers, vegetable waste, etc., then provides substrates for further anaerobic bacterial growth. Local authorities now try to prevent undesirable microbial action in garbage dumps by the methods used in tipping, compacting and draining the garbage so that the landfill can be safely returned to agricultural use or built over. However, trouble due to gas formation, and seepage of waters containing acids and sulphides from old garbage tips, has occurred a number of times in different areas. In some large garbage tips gas has been forming for many years and has accumulated in the tip under a soil covering. This gas can be tapped and used. However, as gas formation is slow the tapping could be more a collection of existing gas than a continuous source. In most cases the amounts of gas in garbage landfills would not be worth collecting, but there are places with very large landfill sites where it seems that commercially useful quantities of biogas are available, and plants have been erected for gas collection in the USA. Some descriptions of these are given in a review by Boyle (1979). The plants are boreholes with suitable interconnecting piping and apparatus for pumping, scrubbing and transferring the gas to the point of use. As the underground microbial action is uncontrolled the gas from different dumps or different areas of a big landfill could vary in methane and impurity (e.g. hydrogen sulphide) content.

The collection of gas from landfill sites can only occur in certain areas where the site is big and the climatic conditions and drainage, etc., on the site have been conducive to gas formation, and selection of suitable sites will be a matter of test boring and calculation. However, as was previously said, the object of modern landfilling is to prevent microbial action which will render the site unfit for further use, so the future of gas from garbage would seem to lie more in the controlled digestion in purpose-built plant rather than in the, somewhat fortuitous, collection of gas from rotting dumps.

A further idea was put forward by Augenstein *et al.* (1976) for landfilling especially for the production of biogas. The sorted garbage would be shredded and mixed with sewage sludge and filled into plastic 'cells' fitted with a gas-tight cover and suitable piping. These cells would be packed into the landfill area and allowed to digest and the gas collected. One would thus have a number of batch digesters, being added to at the rate of perhaps one a month and taken out of commission (the authors suggested on the basis of laboratory tests) after six months to three years, depending on ambient temperature, etc. When gassing ceased the 'cell' would be covered with earth and the land reclaimed for other uses.

REFERENCES

AUGENSTEIN, D. C., WISE, D. L. & WENTWORTH, R. L. (1976). *Resource Recovery and Conservation* **2**, 103.

BOUSFIELD, S., HOBSON, P. N. & SUMMERS, R. (1972). Abstract from ARC Farm Waste Disposal Conf., Glasgow. p. 65.

BOUSFIELD, S., HOBSON, P. N. & SUMMERS, R. (1979). *Ag. Wastes* **1**, 161.

BOYLE, W. C. (1979). Handout from 1st Int. Symp. An. Dig., Cardiff.

CHESSHIRE, M. (1978). In: *Proc. Seminar Anaerobic Digestion of Farm Wastes*, ADAS p. 61.

CHUNG PO (1973). In: *Proc. Int. Biomass Energy Conf.* Winnipeg, Canada. p. XVI-1.

CONVERSE, J. C., EVANS, G. W. & VERHOEVEN, C. R. (1977). Am. Soc. Agric. Eng. Paper No. 77-4051.

COONEY, C. L. & WISE, D. L. (1975). *Biotechnol. Bioeng.* **17**, 1119.

DAVIES, C. (1979). Unpublished communication at a seminar in Aberdeen. Rowett Research Institute, Aberdeen.

DODSON, C. E. (1978). In: *Proc. Seminar Anaerobic Digestion of Farm Wastes*, ADAS p. 65.

DONNELLY, T. (1978). *Proc. Biochem.* **13**, 14.

Farmers Weekly (1979). (November) p. 87.

FINLAY, J. H. (1978). Paper NR/EGMED/G.

GARBER, W. F. (1977). *Prog. Water Technol.* **8**(6), 401.

GARBER, W. F., O'HARA, G. T., COLBAUGH, J. E. & RAKSIT, S. K. (1975). *J. Water Pollution Control Federation* **47**, 950.

GRØN, G. (1980). In: *Anaerobic Digestion* (eds. D. A. Stafford, B. I. Wheatley and D. E. Hughes) Proc. 1st Int. Symp. An. Dig., Cardiff. Applied Science Publishers, London.

HAYES, T. D., JEWELL, W. J., DELL'ORTO, S., FANFONI, K. J., LEUSCHNER, A. P. & SHARMA, D. F. (1980). In: *Anaerobic Digestion* (eds. D. A. Stafford, B. I. Wheatley and D. G. Hughes) Proc. 1st Int. Symp. An. Dig., Cardiff. Applied Science Publishers, London.

HEMENS, J. & SHURBEN, D. G. (1959). *Food Trade Rev.* **29**, 2.

HOBSON, P. N. (1979a). In: *Energy from the Biomass*, Watt Committee on Energy, London, p. 37.

HOBSON, P. N. (1979b). In: *Processes for Chemicals from some Renewable Raw Materials*. Inst. Chem. Eng., London, p. 1.

HOBSON, P. N., BOUSFIELD, S., SUMMERS, R. & MILLS, P. J. (1979). In: *Proc. Engineering Problems with Effluents from Livestock*, an EEC Seminar. p. 492.

HOBSON, P. N. & ROBERTSON, A. M. (1977). *Waste Treatment in Agriculture*, Applied Science Publishers, London.

KHANDELWAL, K. C. (1978). *Compost Sci.* **19**, 22.

KOHN, P. N. (1978). *Chem. Eng.* **85**, 58.

LETTINGA, G., VAN VELSEN, A. F. M., DE ZEEUW, W. & HOMBA, S. W. (1980). In: *Anaerobic Digestion* (eds. D. A. Stafford, B. I. Wheatley and D. E. Hughes) Proc. 1st Int. Symp. An. Dig., Cardiff. Applied Science Publishers, London.

MECKERT, G. W. (1978). Commercial SNG Production from Feedlot Wastes. I.G.T. (August).

MILLS, P. J. (1977). In: *Proc. 9th An. Waste Manag. Conf.*, Cornell Univ.

MORRIS, J. E. (1980). In: *Anaerobic Digestion* (eds. D. A. Stafford, B. I. Wheatley and D. E. Hughes) Proc. 1st Int. Symp. An. Dig., Cardiff. Applied Science Publishers, London.

NEWELL, P. J., COLLERAN, E. & DUNICAN, L. K. (1979). Poster presentation 1st Int. Symp. An. Dig., Cardiff.

NYNS, J. (1978). In: *Proc. Anaerobic Digestion Technology Meeting* ERDA, Ft. Lauderdale, USA.

OBIAS, E. (1976). In: *Proc. Nat. Consultation on Nutrient Recycling Systems*, Los Bacos.

ONO, H. (1965). *Adv. Water Poll. Res.* **2**, 100.

OPENSHAW, K. (1974). *New Scientist* **61**, 271.

PAPOVA, N. M. & BOLOTINA, O. T. (1964). *Adv. Water Poll. Res.* **2**, 97.

PFEFFER, J. T. (1974). *Biotech. Bioeng.* **16**, 771.

PFEFFER, J. T. (1978). *Proc. Biochem.* **13**, 6.

PFEFFER, J. T. (1980). In: *Anaerobic Digestions* (eds. D. A. Stafford, B. I. Wheatley and D. E. Hughes) Proc. 1st Int. Symp. An. Dig., Cardiff. Applied Science Publishers, London.

PYLE, D. L. (1980). In: *Anaerobic Digestion* (eds. D. A. Stafford, B. I. Wheatley and D. E. Hughes) Proc. 1st Int. Symp. An. Dig., Cardiff. Applied Science Publishers, London.

REESEN, L. & STUBE, R. (1978). *Proc. Biochem.* **13**, 21.

ROBERTSON, A. M., BURNETT, G. A., BOUSFIELD, S., HOBSON, P. N. & SUMMERS, R. (1975). In: *Managing Livestock Wastes*. Proc. 3rd Int. Symp. Piggery Wastes. Am. Soc. Agric. Eng. p. 544.

ROSS, W. E. (1954). *Sew. Ind. Wastes* **26**, 140.

RUDOLPHS, W. & TRUBNIK, E. H. (1949). *Sew. Wks. J.* **21**, 1028.

RURAL RESEARCH (1979). CSIRO, Australia. **103**, 8.

SCAMMELL, G. W. (1976). *Ind. Aliment. Agric.* **93**, 169.

SMIL, V. (1977). *Environ.* **19**, 27.

STEFFEN, A. J. (1958). *Biol. Treatment Sew. and Ind. Wastes.* **2**, 126.

STEWART, D. J. (1980). In: *Anaerobic Digestion* (eds. D. A. Stafford, B. I. Wheatley and D. E. Hughes) Proc. 1st Int. Symp. An. Dig., Cardiff. Applied Science Publishers, London.

SWANWICK, J. D. (1975). In: *Methane* (eds. L. Pyle and P. Fraenkel). Intermed. Technol. Publishers, London. p. 5.

TAYLOR, D. W. (1972). In: *Proc. 3rd Nat. Symp. Food Proc. Wastes*, New Orleans, USA.

THOMPSON, L. H. (1977). Paper presented at Eurochem Conf., Birmingham.

WHEATLEY, B. (1978). In: *Proc. Seminar Anaerobic Digestion of Farm Wastes*, ADAS, Reading.

WOOD, B. J., PILLAI, K. R. & RAJARATNAM, J. A. (1979). *Ag. Wastes* **1**, 103.

Appendix 1

PHOTOGRAPHS OF FULL-SCALE WORKING DIGESTERS

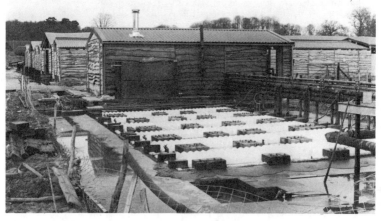

FIG. A.1.   A below-ground digester (227 m³) with floating-top gas-holder and engine and boiler house taking waste from a dairy unit. The gas powers an engine-generator and boiler. Built by Helix Multiprofessional Services, Mortimer Hill, Berks, England.

FIG. A.2.   A 45 m³ concrete digester with independent gas-holder taking mixed waste from an experimental farm. The gas provides hot water for building heating. Built by Davies and Oliver, The Great Yard, Marridge Hill, Ramsbury, Wilts, England.

FIG. A.3. A 400 m³ digester taking piggery waste and fabricated from a commercial glass-lined slurry tank (Smith-Harvestore, Eye, Suffolk, England) and independent gas-holder. The gas fuels an engine-generator. Built by Farm Gas Ltd, Industrial Estate, Bishops Castle, Salop, England and Dale Electrics, Filey, Yorks, England.

FIG. A.4. A 1200 m³ glass-lined steel-tank digester and (foreground) water-sealed gas-holder taking waste from a piggery. The gas powers three engine-generators, part of an 'Anox' complete waste-water treatment unit producing recyclable water and dried solids. Built by D. Evers & Associates, 43 Forgate Street, Worcester, England.

FIG. A.5. A steel-tank digester with a flexible-rubber-top gas-holder in fully-inflated state, taking piggery waste. The gas is used to heat a greenhouse. Built by Kilbees Slurry Digesters Ltd, Winkfield, Windsor Forest, Berks, England.

FIG. A.6. A 'package-unit' digester for small-community, or similar, sewage treatment. Built by Hamworthy Engineering Ltd, Fleets Corner, Poole, Dorset, England.

FIG. A.7. A 3000 m³ contact digester treating 700 m³ of effluent from a wheat-starch factory per day. The gas is used in the starch-drying plant. Built by Biomechanics Ltd, Caxton House, Wellesley Road, Ashford, Kent, England.

FIG. A.8. An 80 m³ digester for sewage treatment similar to the steel-tank farm digesters. Built by Farm Gas Ltd, Industrial Estate, Bishops Castle, Salop, England and Severn and Trent Water Authority, Mapperley Hall, Mapperley, Notts., England.

Appendix 2

SOME ESTIMATES OF WASTES AVAILABLE FOR BIOGAS OR OTHER FUEL PRODUCTION

(All quantities are in tonnes per year)

Crop residues
  Cereals
    Burma   12 388 000   (FAO, 1979)
    Canada   44 906 000   (Kanoksing & Lapp, 1975)
    England and Wales
      (1978)   10 000 000   (MAFF, 1978)
    Thailand   22 135 600   (Kanoksing & Lapp, 1975)
    USA (1973)   546 588 000   (Detnoy & Hesseltine, 1978)
  Millet straw:
    Senegal   1 000 000   (FAO, 1979)
  Rice straw:
    Korea (Republic)   10 000 000   (FAO, 1979)
  Maize (stalks and cobs):
    Nepal   6 800 000   (FAO, 1979)

Other residues
  Coconut:
    Philippines   4 300 000   (FAO, 1979)
  Fruit and vegetables:
    Banana (leaves and stems)
    Somalia   800 000   (FAO, 1979)
    Canning and citrus peel
      extraction
    Peru   2 800   (FAO, 1979)
  Legumes:
    Burma   1 500 000   (FAO, 1979)
  Oil palm:
    Malaysia   9 840 000   (FAO, 1979)

Industrial wastes
  Brewery:
    Yeast waste
      Fiji   1 000   (FAO, 1979)
  Sugar:
    Bagasse
      Burma   800 000   (FAO, 1979)
      Columbia   4 000 000   (FAO, 1979)
      Cuba   22 000 000   (FAO, 1979)
      Guyana   1 800 000   (FAO, 1979)
    Filter mud
      Central America and
        Panama   520 000   (FAO, 1979)
    Sugar beet residue
      UK (by 2000)   600 000   (Langley, 1980)

Domestic refuse
      London, UK (1971)   2 858 000   (GLC, 1972)
      USA (1980)   160 000 000   (Boyle, 1979)

Farm animal wastes
  Cattle:
      [a]UK (by 2000)   7 000 000   (Langley, 1980)
      USA (1973)   195 000 000   (Detnoy & Hesseltine, 1978)
      Japan (1977)   35 473 000   (Haga *et al.*, 1979)
  Pigs:
      England and Wales
        (1977)   11 100 000   (RCEP, 1979)
      [a]UK (by 2000)   1 000 000   (Langley, 1980)
      USA (1973)   22 400 000   (Detnoy & Hesseltine, 1978)
      Japan (1977)   12 121 000   (Haga *et al.*, 1979)
  Poultry:
      England and Wales
        (1977)   4 600 000   (RCEP, 1979)
      Guatemala   1 400   (FAO, 1979)
      Mauritius   2 700   (FAO, 1979)
    Feather and offal   700   (FAO, 1979)
      Japan (1977)   8 897 000   (Haga *et al.*, 1979)
      USA (1973)   4 590 000   (Detnoy & Hesseltine, 1978)
      [a]UK (by 2000)   800 000   (Langley, 1980)

[a] Calculated as dry and ash-free. Other quantities are as produced.

General manure:

Korea (Republic)	12 000 000	(FAO, 1979)
Philippines	12 700 000	(FAO, 1979)

REFERENCES

BOYLE, W. C. (1979). Energy recovery from sanitary landfills—a review.

DETNOY, R. W. & HESSELTINE, C. W. (1978). *Proc. Biochem.* **13**(9), 2–9. Availability and utilisation of agricultural and agroindustrial wastes.

FAO (1979). Agricultural residues: quantitative survey. FAO, Rome.

GLC (1972). Ann. Rept. Dept. Publ. Hlth. Eng. GLC, London.

HAGA, K., TANAKA, H. & HIGAKI, S. (1979). *Ag. Wastes* **1**(1), 45–57. Methane production from animal wastes and its prospects in Japan.

KANOKSING, P. & LAPP, H. M. (1975). Paper presented at Can. Soc. Ag. Eng. Ann Meeting, Brandon Univ., Brandon, Manitoba, Canada. Feasibility for energy recovery from cereal crop wastes.

LANGLEY, K. (1980). Renewable energy through anaerobic digestion. In: *Anaerobic Digestion* (eds. D. A. Stafford, B. I. Wheatley and D. E. Hughes) Proc. 1st Int. Symp. An. Dig., Cardiff. Applied Science Publishers, London.

MAFF (1978). Report of Straw Utilisation Conference. Ministry of Agriculture, Fisheries and Food.

Royal Commission on Environmental Pollution (RCEP) (1979). Seventh report. *Agriculture and pollution.* HMSO, London.

Appendix 3

GLOSSARY OF TERMS

Pollution and Pollution Control

BOD, Biochemical Oxygen Demand. The oxygen uptake by bacteria growing in the water determined under defined conditions (see e.g. *Standard methods*, 1965). A measure of the material, mostly in solution, liable to cause immediate pollution of a water course or land. Usual units mg/litre.

COD, Chemical Oxygen Demand. Oxygen uptake in a chemical oxidation of the waste (solids and solutes). A measure of the material liable to cause long-term, as well as immediate, pollution. Usual units mg/litre.

TS, Total Solids. The material left after evaporation of water from a sample. Usual units % (wt/vol) or mg/litre.

VS, Volatile Solids. The material burnt off on incineration of the TS, i.e. the organic material in the sample. Usual units % of TS.

SS or VSS, Suspended, or Volatile Suspended, Solids. Total, or organic, material in suspension, not solution, in the sample. Usual units % (wt/vol) or mg/litre.

VFA, Volatile Fatty Acids. Acids of the type acetic in solution in the sample. Usual units mg/litre, measured as acetic acid (see later).

NH_3N, Ammonia nitrogen. Total ammonia in solution in the sample. Measured as ammonia but usually reported as ammonia nitrogen (i.e. 14/17 wt, of ammonia) in mg/litre.

Alkalinity. A measure of the capacity of a liquid to resist the establishment of alkaline conditions. Expressed as the concentration of calcium carbonate which would have the equivalent capacity to neutralise strong acids.

Metal ions, chloride, sulphate, etc., are usually reported in mg/litre.

Some Microbiological and Chemical Terms

Bacterium. Single-cell organism reproducing by division into two cells. Most bacteria in anaerobic digesters are about 1–3 μm in length or diameter. The organisms can be rod-like, straight or with single or multiple curvatures, or nearly spherical (cocci). As they divide the bacterial cells may separate or remain attached to form chains or groups of two or more cells. Bacteria can, and many always do, live suspended in a nutrient medium, and some are capable of directed movement. However, bacteria degrading and utilising solid substrates are generally attached to the solid and form groups of 'microcolonies' on or in cavities in the solid as they grow and divide. Bacteria can be *aerobic*, i.e. they require oxygen for growth and obtain energy for growth by conversion of an 'energy source' with oxygen to carbon dioxide and water as do higher animals; or they can be *anaerobic*, i.e. they can and must live in an atmosphere devoid of oxygen and obtain energy by reactions, such as formation of acids or gases, in which oxygen does not play a part. *Facultative* or *facultatively anaerobic* bacteria can utilise either pathway and can live with or without oxygen. It should be noted that as bacteria live in water (even on an apparently dry material) it is oxygen in solution that is used; the atmosphere above the medium serves only to replenish the oxygen in solution as it is used up by the bacteria. Similarly methanogenic bacteria use hydrogen and carbon dioxide in solution and the gases are replenished from the atmosphere above the medium or by formation by other bacteria suspended in the medium. *Gram-positive* and *Gram-negative* are two broad classifications of bacteria based on their ability to retain a purple dye added by a method first described by Gram. The presence or absence of the stain is determined by microscopic observation of the bacteria.

Energy sources. Sometimes referred to generally as 'Carbon Source'. A substance, often a carbohydrate, but always a carbon compound which can be metabolised to give energy for synthesising new cell substances, recycling cell substances, transport of materials into the cells, locomotion, etc. A proportion of the weight of energy source supplied may be used after fragmentation to build up cell substances.

Nitrogen source. Organic or inorganic (e.g. ammonia) compounds containing nitrogen and used to form the nitrogenous material of the bacterial cells. Some bacteria can build up all the cell materials from a simple source of nitrogen such as ammonia; others require a complex organic source with many of the amino acids and other cell substances present as complete compounds.

Medium or Culture Medium. An aqueous solution, although some constituents may be in suspension, of the energy and nitrogen sources, sulphur compounds, salts, vitamins or other materials required for growth of a particular bacterium or a number of different kinds of bacteria. 'Medium' usually refers to the laboratory product, but materials such as sewage waters may also be referred to as a medium. The solution and suspension may be, in the laboratory, solidified or semisolidified by addition of a gelling agent such as agar.

Carbohydrate. A compound made up from carbon, hydrogen and oxygen atoms and in the context of the present book generally a 'polymer', i.e. a large molecule made up from a number (often hundreds) of smaller molecules joined together in a particular way. Cellulose is a polymer of the simple sugar molecule, glucose, as is starch, but the linkage between the glucose molecules differs in the two polymers, conferring different properties on the polymers. Cellulose molecules aggregate and form fibres which do not disperse or dissolve in hot, or cold, water. Starch molecules form a more open structure which can be fragmented and dispersed in hot water. 'Hemicellulose' is a name given to another type of carbohydrate, mainly a polymer of the sugar xylose, which has a different structure from glucose. 'Pectins' are essentially polymers of the sugar acid, galacturonic acid. 'Fructosans' are polymers of another sugar, fructose.

Sugar (*simple sugar*). In this text it refers to a 'monosaccharide', a unit such as glucose from one of the above polymers, or a 'disaccharide', two such units joined together. Some sugars occur as such in nature; e.g. sucrose (cane or beet sugar) a disaccharide of glucose plus fructose. Molasses is essentially a solution of sucrose.

Protein. A polymeric molecule made up from monomer units of 'amino acids', but while polymeric carbohydrates are generally made up from one monomer, proteins contain a number of different kinds of amino acid linking in an array which repeats itself to give a molecule with many hundreds of amino acids. Amino acids are organic acids, in many cases basically the same as the volatile fatty acids, but with an amino group of nitrogen and hydrogen ($—NH_2$; cf. formula for ammonia, NH_3) added to the acid molecule.

Non-protein nitrogen (*NPN*). Nitrogen contained in materials other than proteins. In digesters it is principally ammonia, but bacteria and vegetable materials and urines contain other NPN sources. These can be simple compounds such as urea, or much more complex organic compounds such as purines and pyrimidines, carbohydrates containing nitrogen atoms and so on. Amino acids not combined into protein molecules also come under this heading.

Organic acids. *Volatile fatty acids* (*VFA*) are a group of acids including formic acid which has one carbon atom (structure HCOOH, the COOH being the acidic group) and acetic acid, which has two carbon atoms (structure CH_3COOH (vinegar)). Propionic acid has three carbon atoms (CH_3CH_2COOH), butyric acid four, etc. Lactic and succinic are also organic acids, but with a more complex structure. *Long-chain fatty acids* are of the same class as the volatile fatty acid but have many more carbon atoms, e.g. stearic acid has 18 carbon atoms in the chain. These acids are constituents of the fat tissue of animals.

pH. A measure of the acidity or alkalinity of a solution. A pH of 7 is neutral, a pH below 7 is acidic, above 7 is alkaline. The extremes of the range (pH 1 and 14) could be represented by solutions of sulphuric acid and sodium hydroxide (caustic soda).

Eh or oxidation–reduction potential. In the present context a measure to some extent of the amount of oxygen in solution in a liquid. Determined by using a hydrogen electrode, but for convenience in practice measured by the voltage difference between a platinum electrode and a standard calomel electrode, with a correction for the voltage of the calomel electrode. A positive reading of say 200 mV or more is given in water partially saturated with oxygen, a value of zero in water with little oxygen, and values of -200 or more, at which the anaerobic bacteria grow, in water with no detectable oxygen in solution.

Detention, or Retention, Time. The time for which a particular portion of the input to a digester or other continuous culture system remains in the digester vessel (volume of digester divided by flow rate). Units, days or hours.

Dilution rate. The reciprocal of detention time, flow rate divided by volume. Units, reciprocal time, 1/days or 1/hours.

Index